Fermentation Microbiology and Biotechnology

Edited by
Dr Mansi El-Mansi
and
Professor Charlie Bryce

TAYLOR & FRANCIS GROUP
ALERE FLAMMAM
·1798 – 1998·

UK   Taylor & Francis Ltd, 11 New Fetter Lane, London EC4P 4EE
USA   Taylor & Francis Inc., 325 Chestnut Street, 8th Floor, Philadelphia, PA 19106, USA

Taylor & Francis is an imprint of the Taylor & Francis Group

British Library Cataloguing in Publication Data

A catalogue record for this book is available from the British Library.
ISBN 0–7484–0733–2

Library of Congress Cataloguing in Publication Data are available

Typeset in 10/12 Sabon by Graphicraft Limited, Hong Kong
Printed by T.J. International Ltd, Padstow, UK

Dedication

I wish to dedicate my contributions, with affection and gratitude, to
my wife Gillian for her endurance and to
my sons, Adam and Sammy, for their love and
the much needed happiness which they brought into my life.

Mansi El-Mansi

Contents

Contributors

A.R. Allman
Infors UK Ltd, Fortune House, 10 Bridgeman Terrace, Wigan
WN1 1SX, UK

H. Blachere
Inceltech, 31100 Toulouse, France

C.F.A. Bryce
Department of Biological Sciences, Napier University, Edinburgh
EH10 5DT, Scotland, UK

M.F. Cardosi
Department of Biological Science, University of Paisley, Paisley
PA1 2BE, UK

G. Corrieu
Laboratoire de Genie et de Microbiologie des Procedes
Alimentaires-INRA, Centre de Biotechnologie Agro-Industrielle,
78850 Thiverval, Grignon, France

B. Dahhou
L.A.A.S. du C.N.R.S., 7 avenue du Colonel Roche, 31077 Toulouse Cedex, France

E.M.T. El-Mansi
Applied Microbiology and Biotechnology Group, Department
of Microbiological Sciences, Napier University, Edinburgh EH10
5DT, Scotland, UK

M. Fischer
Bioflux Ltd, Colville Building, 48 North Portland Street, Glasgow
G1 1XM, UK

C.J.L. Gershater
SmithKline Beecham Pharmaceutical, New Frontiers Science Park,
Essex, UK

G. Goma
Centre de Biologiniérie Gilbert Durand, INSA, Avenue de
Rangeuil, 31077 Toulouse Cedex, France

B.S. Hartley
Grove Cottage, Elsworth, Cambridge CB3 8HP, UK

I.S. Hunter
Department of Pharmaceutical Sciences, University of Strathclyde, Glasgow G1 1XW, UK

E. Latrille
Laboratoire de Genie et de Microbiologie des Procedes Alimentaires-INRA, 78850 Thiverval, Grignon, France

J.C. Melville
Bioflux Ltd, Colville Building, 48 North Portland Street, Glasgow G1 1XM, UK

D.M. Mousdale
Bioflux Ltd, Colville Building, 48 North Portland Street, Glasgow G1 1XM, UK

J. Nielsen
Centre for Process Biotechnology, Technical University of Denmark, Denmark

G. Roux
L.A.A.S. du C.N.R.S., 7 avenue du Colonel Roche, 31077 Toulouse Cedex, France

G. Stephanopoulos
Department of Chemical Engineering, Massachusetts Institute of Technology, Cambridge, MA, USA

J.-P. Steyer
Research Group in Process Modelling, Control and Diagnosis, Laboratoire de Biotechnologie de l'Environnement, Avenue des Etangs, INRA, 11100 Narbonne, France

Preface

Biotechnology has been defined as any technique that uses living organisms, or substances from those organisms, to make or modify a product, to improve plants or animals, or to develop microorganisms for specific uses. It is clear from this broad, generic definition that biotechnology need not necessarily be considered a modern practice, although it is generally recognized that in recent years this discipline has witnessed exciting advancements.

As a consequence of these exciting developments, the principal aim of this book is to illustrate the current thinking and the frontline research in the area of fermentation biotechnology, paying special attention to key application areas and novel support technologies. To this end, it was pleasing to be able to persuade a group of recognized experts from both university and industry each to contribute a chapter dealing with their respective expertise whilst, at the same time, creating a valuable collection of related, dependent and interwoven accounts that will serve the needs of current and future workers in this exciting and growing area of work and help to raise the competitive edge of industrial production.

The nature and scope of this text can be inferred from scanning the individual chapter headings, from which it can be seen that the reader is introduced to the design, operation and applications of fermentors, taken through the basic principles and the relevant applications of fermentation kinetics, shown the need for effective bioprocess monitoring through the use of specific biosensors, introduced to the concepts of metabolic flux analysis and metabolic engineering (which are generally considered to be underestimated aspects of fermentation biotechnology) and finally shown the value of effective control systems and strategies from conventional to current state-of-the-art knowledge-based and expert systems, neural networks and fuzzy logic, genetic algorithms and modelling, and ending with a fully developed operating and supervision system.

In this way it is hoped that the reader will benefit significantly from a comprehensive and comprehensible account of the entire fermentation process, the underpinning science and technology, the efficient and effective monitoring and control of the process with the ultimate goal of generating end-products which contribute to the improvement of the quality of our lives.

The Editors
May 1999, Edinburgh

1 Fermentation Biotechnology: An Historical Perspective

E.M.T. El-Mansi, C.F.A. Bryce and B.S. Hartley

1.1 Fermentation: an ancient tradition

Fermentation has been known and practised by mankind since prehistoric times, long before the underlying scientific principles were understood. That such a useful technology should arise by accident will come as no surprise to those people who live in tropical and subtropical regions; where, as Marjory Stephenson put it, 'every sandstorm is followed by a spate of fermentation in the cooking-pot'. For example, the production of bread, beer, vinegar, yoghurt, cheese and wine were well established technologies in ancient Egypt (Figures 1.1 and 1.2). It is an interest-

---

Figure 1.1
Bread making as depicted on the wall of an ancient Egyptian tomb, c. 1400 BC. (Reprinted with the kind permission of the Fitzwilliam Museum, Cambridge, England.)

Figure 1.2
Grape treading and wine making as depicted on the walls of Nakhte's tomb, Thebes, c. 1400 BC. (Reprinted with the kind permission of AKG, London, England/Erich Lessing.)

ing fact that archaeological studies have revealed that bread and beer, in that order, were the two most abundant components in the diet of ancient Egyptians. Everyone, from the Pharaoh to the peasant, drank beer for social as well as ritual reasons. Archaeological evidence has also revealed that ancient Egyptians were fully aware not only of the need to malt the barley or the emmer wheat but also of the need for starter cultures, which at the time may have contained lactic acid bacteria in addition to yeast.

1.2   The rise of fermentation microbiology

With the advent of the science of microbiology, and in particular fermentation microbiology, we can now shed light on these ancient and traditional activities. Consider for example the age-old technology of wine making, which relies upon crushing the grapes (Figure 1.2) and letting Nature takes its course, i.e. fermentation. Many microorganisms can grow on the grape sugars more readily and efficiently than yeasts, but few can withstand the osmotic pressure arising from the high sugar concentrations. Also, as sugar is fermented, the alcohol concentration rises to a level at which only the osmo-tolerant, alcohol-tolerant cells can survive. Hence men of ancient civilizations did not need to be skilled microbiologists in order to enjoy the fruits of this popular branch of fermentation microbiology.

In fact, the scientific understanding of fermentation microbiology and, in turn, biotechnology only began in the 1850s, after Louis Pasteur had succeeded in isolating two different forms of amyl alcohol, of which one was optically active (L, laevorotatory) while the other was not. Rather unexpectedly, the optically inactive form resisted all Pasteur's attempts to resolve it into its two main isomers, i.e. the laevorotatory (L) and the dextrorotatory (D) forms. It was this observation that led Pasteur into the study of fermentation, in the hope of unravelling the underlying reasons behind his observation, which was contrary to current understanding of stereochemistry and crystallography.

In 1857, Pasteur published the results of his studies and concluded that fermentation is associated with the life and structural integrity of the yeast cells rather than with their death and decay. He reiterated the view that the yeast cell is a living organism and that the fermentation process is essential for the reproduction and survival of the cell. In his paper the words 'cell' and 'ferment' are used interchangeably, i.e. the yeast cell is the ferment. The publication of this classic paper marks the birth of fermentation microbiology, and in turn biotechnology, as a new scientific discipline. Guided by his knowledge and armed with his experimental observations, Pasteur was able to confidently challenge and reject Liebig's idea that fermentation occurs as a result of contact with decaying matter. He also ignored the well documented view that fermentation occurs as a result of 'contact catalysis' though it is possible that this was not suspect in his view. The term 'contact catalysis' probably implied that fermentation is brought about by a chain of enzyme-catalysed reactions, and we might also credit Pasteur with the concept of an enzyme as 'in yeast'.

Although Pasteur's interpretations were essentially physiological rather than biochemical, they were pragmatically correct. During the course of his further studies, Pasteur was also able to establish not only that alcohol was produced by yeast through fermentation but also that souring was a consequence of contamination with bacteria that were capable of converting alcohol to acetic acid. Souring could be avoided by heat treatment at a certain temperature for a given length of time. This eliminates the bacteria without adversely affecting the organoleptic qualities of the beer or the wine, a process we now know as **pasteurization**.

A second stage in the development of fermentation microbiology and biotechnology, more collectively known as fermentation biotechnology, began in 1877, when Moritz Traube proposed the theory that fermentation and other chemical reactions are catalysed by protein-like substances and that, in his view, these substances remain unchanged at the end of the reactions. Furthermore, he described fermentation as a sequence of events in which oxygen is transferred from one part of the sugar molecule to another, culminating in the formation of a highly oxidized product, i.e. CO_2, and a highly reduced product, i.e. alcohol. Considering the limited knowledge of biochemistry in general and enzymology in particular at the time, Traube's remarkable vision was to prove some 50 years ahead of its time.

In 1897, two years after Pasteur died, Eduard Buchner was successful in preparing a cell-free extract which fermented sugar. This discovery was received with a great deal of interest, not only because it was the first evidence of fermentation without a living organism but also because it was in sharp contrast to the theory proposed by Pasteur. In the early 1900s, the views of Pasteur were modified and extended to stress the idea that fermentation is a function of a living, but not necessarily multiplying, cell and that it is not a single step but rather a chain of different reactions, each of which is probably catalysed by a different enzyme.

1.3   Developments in metabolic and biochemical engineering

Outbreak of the First World War provided an impetus and a challenge to produce certain chemicals which, for one reason or another, could not be manufactured by conventional means. For example, there was a need for glycerol, an essential component in the manufacture of ammunition, because no vegetable oils could be imported due to the naval blockade. German biochemists and engineers were able to adapt yeast fermentation, turning sugars into glycerol rather than alcohol. Although this process enabled the Germans to produce in excess of 100 tonnes of glycerol per month, it was abandoned as soon as the war was over because glycerol could be made very cheaply as a by-product of the soap industry. There was also, of course, a dramatic drop in the level of manufacture of explosives and, in turn, the need for glycerol.

The diversion of carbon flow from alcohol production to glycerol formation was achieved by adding sodium bisulphite, which forms an adduct with acetaldehyde (Figure 1.3), thus preventing it from being reduced to ethanol. Consequently, NADH accumulates intracellularly, thus perturbing the steady-state redox balance (NAD^+/NADH ratio) of the cell. The drop in the intracellular level of NAD^+ is accompanied by a sharp drop in the flux through glyceraldehyde-3-phosphate dehydrogenase which, in turn, allows the accumulation of the two

Figure 1.3

The metabolic network of alcohol fermentation and the role of sodium bisulphite in the diversion of carbon flow from alcohol production to glycerol formation. Note that bisulphite arrests acetaldehyde and that the resulting adduct is not a substrate for alcohol dehydrogenase. The redox balance, i.e. the NAD$^+$/NADH ratio, is therefore perturbed and to redress such imbalance the cells have to divert carbon flow (dashed route) towards the reduction of dihydroxyacetone-3-phosphate to glycerol-3-phosphate, with the concomitant regeneration of NAD$^+$. The glycerol-3-phosphate thus generated is then dephosphorylated to glycerol.

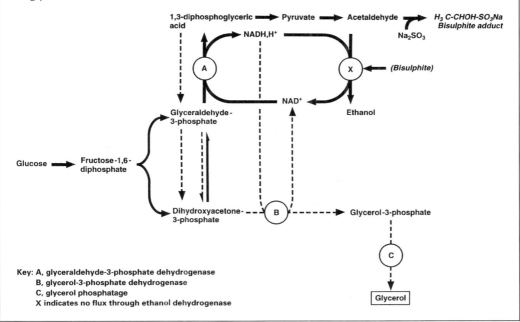

isomeric forms of triose phosphate, i.e. glyceraldehyde-3-phosphate and dihydroxyacetone-3-phosphate. Accumulation of the latter, plus high intracellular levels of NADH, induce the expression and in turn flux through glycerol-3-phosphate dehydrogenase, thus restoring the redox balance within the cell and diverting catabolic flux to glycerol formation instead of ethanol (Figure 1.3). Although this explanation is with the hindsight of modern biochemistry, the process can be viewed as an early example of **metabolic engineering**.

Following the First World War, research into yeast fermentation was largely influenced by the work of Neuberg and his proposed scheme (**biochemical pathway**) for the conversion of sugars to alcohol (**alcohol fermentation**). Although Neuberg's scheme was far from perfect and proved erroneous in many ways, it provided the impetus and the framework for many scientists at the Delft Institute, who vigorously pursued research into oxidation/reduction mechanisms and the kinetics of product formation in a wide range of enzyme-catalysed reactions. Such studies were to prove important in the development of modern biochemistry as well as fermentation biotechnology.

While glycerol fermentation was abandoned immediately after the First World War, the acetone–butanol fermentation process, catalysed by *Clostridium acetobutylicum*, flourished. Production lines were modified to accommodate the new approach of 'Fill and Spill', which permitted substantial savings in fuels without adversely affecting the output of solvent production during the course of the Second World War. However, as soon as the production of organic solvents as a by-product of the petrochemical industry became economically viable, the acetone–butanol fermentation process was discontinued.

1.4   Discovery of antibiotics and genetic engineering

The discovery of penicillin and its antibacterial properties in the early 1940s represents a landmark in the development of modern fermentation biotechnology. This discovery, to a country at war, was both sensational and invaluable. However, *Penicillium notatum*, the producing organism, was found to be susceptible to contamination by other organisms and so aseptic conditions were called for. Such a need led to the introduction of so-called **stirred tank bioreactors** which minimize contamination with other unwanted organisms. The demand for penicillin prompted a worldwide screen for alternative penicillin-producing strains, leading to the isolation of *Penicillium chrysogenum* which produces more penicillin than the original isolate, *P. notatum*. *Penicillium chrysogenum* was then subjected to a very intensive programme of random mutagenesis and screening. Mutants which showed high levels of penicillin production were selected and subjected to further round of mutagenesis, and so on. This approach was successful, as indicated by the massive increase in productivity from less than 1 gl^{-1} to slightly over 20 gl^{-1} of culture.

Once the antibacterial spectrum of penicillin was determined and found to be far from universal, pharmaceutical companies began the search for other substances with antibacterial activity. This screening programme led to the discovery of *Actinomycetes* and the many antibacterial agents produced by various members of this genus of filamentous bacteria. Although the search for new antibiotics is never over, intensive research programmes involving the use of genetic and metabolic engineering were initiated with the aim of increasing the productivity and potency of current antibiotics. For example, the use of genetic and metabolic engineering has increased the yield of penicillin by 100%, from 20 g l^{-1} to 40 g l^{-1}.

1.5 The rise and fall of single cell protein

The latter part of the 1960s saw the rise and fall of **single cell protein** (SCP) production from oil or natural gas. A large market for SCP was forecast as the population in the third world, the so-called under-developed countries, continued to increase despite a considerable shortfall in food supply. However, the development of SCP died in its infancy, largely due to the sharp rise in the price of oil which made it economically non-viable. Furthermore, improvements in the quality and the yields of traditional crops did not help the cause of SCP production.

1.6 Fermentation biotechnology and the production of amino acids

The next stage in the development of fermentation biotechnology was dominated by the success in the use of regulatory control mechanisms for the production of amino acids. The first breakthrough came in the late 1950s and early 1960s, when a number of Japanese researchers discovered that regulatory mutants, isolated by virtue of their ability to resist amino acid analogues, were capable of over-producing amino acids. The exploitation of such a discovery, however, was hampered by the induction of degradative enzymes once the extra-cellular concentration of the amino acid increases beyond a certain level, e.g. accumulation of tryptophan induces the production of tryptophanase thus initiating the breakdown of the amino acid. This problem was resolved by the use of penicillin which, with tryptophan as the sole source of carbon in the medium, eliminates the growing cells, i.e. those capable of metabolizing tryptophan, but not those which are quiescent. Following the addition of penicillin, the mutants which had survived the treatment (3×10^{-4}) were further tested and enzymic analysis revealed that one mutant was totally devoid of tryptophanase activity. This approach was soon extended to cover the production of glutamate and other amino acids, particularly those which are not found in sufficient quantities in plant proteins. The successful use of regulatory mutants stimulated interest in the use of auxotrophic mutants for the production of other chemicals. The rationale is that auxotrophic mutants will negate feedback inhibition mechanisms and in turn allow the accumulation of desired end-product. For example, an arginine-auxotroph was successfully used in the production of ornithine while a homoserine-auxotroph was used for the production of lysine.

1.7 Fermentation biotechnology: future prospects

Today the pace of progress in fermentation biotechnology is fast and furious, particularly since the advent of genetic engineering and the recent advances in computer sciences and process control. The subject currently encompasses a host of different scientific disciplines, including microbiology, biochemistry, molecular genetics, bioinformatics and, last but not least, engineering. In this book we hope to address the multidisciplinary nature of this subject to highlight its many fascinating aspects and to provide a stepping stone in its progress. As we enter a new era of scientific progress, in which the use of renewable resources for the production of desirable end-products is recognized as an urgent need, fermentation biotechnology has a major role.

2 Fermentors: Design, Operation and Applications

Anthony R. Allman

2.1 Batch culture fermentation

The purpose of this chapter is to give an introduction to the methodology of batch fermentation using a small, autoclavable, bench-top fermentor system. Specifically, this chapter should provide an understanding of the following:

- the engineering and biological concepts of fermentors;
- the principles behind the instrumentation used with a modern fermentor;
- the assembly and preparation of a fermentor vessel and ancillary equipment for autoclaving;
- how samples are taken from the fermentor vessel during operation;
- how simple continuous culture can be accomplished.

A **fermentor** is a system consisting of a few pieces of equipment which provide controlled environmental conditions for the growth of microbes (and/or production of specific metabolites) in liquid culture whilst preventing entry and growth of contaminating microbes from the outside environment.

2.2 The main components of a fermentor and their uses

The main subdivisions of a bench-scale fermentor are as follows:

- base components including drive motor, heaters, pumps, gas control, etc.;
- vessel and accessories;
- peripheral equipment such as reagent bottles;
- instrumentation and sensors.

These components combine to perform the following functions:

- provide operation free from contamination;
- maintain a specific temperature;
- provide adequate mixing and aeration;
- control the pH of the culture;
- allow monitoring and/or control of dissolved oxygen;
- allow feeding of nutrient solutions and reagents;
- provide access points for inoculation and sampling;
- use fittings and geometry relevant to scale-up;
- minimize liquid loss from the vessel;
- facilitate the growth of a wide range of organisms.

CONTINUOUS CULTURE
A method of allowing culture to be grown in a fermentor at a specific growth rate. The growth rate is determined by the flow of medium through the vessel (dilution rate).

The example used to illustrate these points is the smallest, simplest system which displays all these components and attributes, i.e. a small, autoclavable, bench-top fermentor. Mention is made of larger, *in situ* sterilizable units and special features which can be added to achieve certain objectives such as containment work. Also, a range of basic applications are illustrated, e.g. mammalian cell culture as a starting point for anyone contemplating the use of small-scale fermentors for these purposes. The specific information provided should be treated with a great deal of circumspection as the parameter values required for any particular microbes or cell line could be completely different.

2.3   Component parts of a 'typical' vessel

The vessel is constructed either as a single-walled cylinder of borosilicate glass with a flat bottom or as a glass-jacketed system which typically has a round bottom. The top plate is made from '316' stainless steel and is compressed onto the vessel flange by an easily released clamping system. A seal separates the vessel glass from the top plate. Port fittings of various sizes are provided for insertion of probes, inlet pipes, exit gas cooler, cold fingers, sample pipes, etc. These work by compressing the sides of the probe/pipe against an O-ring seal. A special inoculation port will have a membrane seal held in place with a collar. Culture can be withdrawn into a sampling device or a reservoir bottle via a sample pipe situated in the bulk of the fermentor fluid. A gas sparger is also fixed into the top plate and this terminates in a special assembly which ensures that incoming air is dispersed efficiently within the culture by the flat-bladed 'Rushton-type' impellors fixed to the drive shaft. A drive motor provides stirring power to the drive shaft and is usually fitted directly to the drive hub on the vessel top plate. An exit gas cooler works like a simple Liebig condenser to remove as much moisture as possible from the gas leaving the fermentor to prevent excessive liquid losses during the fermentation and wetting of the exit air filter.

A narrow platinum resistance (Pt-100) temperature sensor completes the list of minimum essential fittings. Temperature control is either by direct heating using a heater pad or by circulating warm water around the vessel jacket. If direct heating is used, a cold finger is used to cool the vessel contents.

PT-100 TEMPERATURE SENSOR
A platinum resistance electrode used to give an accurate indication of vessel temperature by relating changes in electrical resistance of the sensor to temperature.

The sensors which fit into the vessel do so either by direct coupling with a thread on the body of the electrode, as is the case in the gel-filled type of pH electrode, or by a special fitting on the vessel top plate, as in the case of the dissolved oxygen electrode (polarographic type). Another system involves the use of a simple compression fitting which holds the body of the

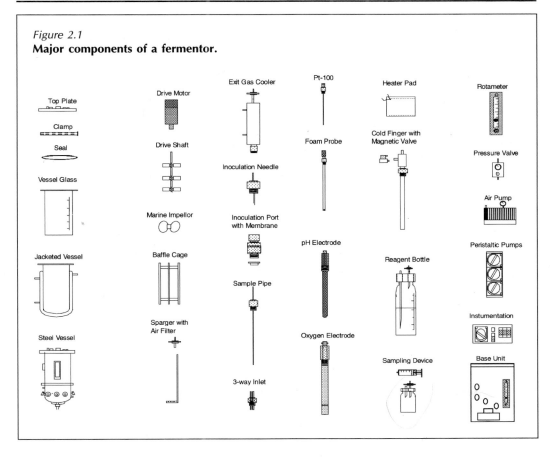

Figure 2.1
Major components of a fermentor.

COLD FINGER
A closed pipe or coil which passes through the fermentor top plate and allows cooling water to circulate, to act as a heat exchanger with the culture.

electrode, as with the foam probe. In this case, the height is variable and the tip of the probe is above the culture fluid. Figure 2.1 illustrates all the main components of a fermentor.

2.4 Peripheral parts and accessories

2.4.1 Reagent pumps

Pumps are normally part of the instrumentation system for pH and antifoam and control. Peristaltic pumps are used and the flow rate is usually fixed with a timed 'shot and delay' feed system of control. Flow rates will depend on the bore of the tubing used and it is a good idea to move the tubing every few days so that one piece does not become worn out by the pump rollers. The peristaltic tubing links the reservoir bottles with the vessel multi-way inlet. The tubing is clamped shut during autoclaving, already connected to the vessel, and opened for active addition of reagent. Alternatively, the bottles can be connected following autoclaving by using an aseptic coupling to join the tubing between the pump and the multi-way inlet. A dip tube goes to the bottom of the bottle. A filter in the top of the reservoir bottle is kept open during autoclaving to prevent a build-up of pressure. For extremely high accuracy of addition, the reservoir bottles are placed on analytical balances which can be used to determine how much reagent has been pumped in a given time.

2.4.2 Medium feed pumps and reservoir bottles

Medium feed pumps are often variable speed to give the maximum possible range of feed rates. The speed of operation of the pump can be set manually or the whole system put under computer control. The reservoir bottles are usually larger, e.g. 5–20 l, but are prepared in the same way as normal reagent bottles. These bottles may have to be changed several times during a long continuous culture experiment, so it is usual to fit an aseptic coupling in the tubing. An effluent pump is often used to remove culture fluid from the fermentor vessel into a storage reservoir bottle.

2.4.3 Rotameter/gas supply

Gas input, usually air, can be provided by use of a laboratory air supply or from a separate pump (which *must* be oil-free). A variable area flow meter or **rotameter** is used to control the air flow rate into the fermentor vessel. A pressure regulator valve before the rotameter ensures safer operation. A sterile filter (usually 0.22 µm) is used as a bridge between the tubing from the rotameter and that connected to the air sparger of the fermentor. A second filter on the exit gas cooler stops microbes being released into the laboratory air as the gas leaves the fermentor under a slight positive pressure.

2.4.4 Sampling device

This allows culture fluid to be removed aseptically during the fermentation at intervals decided by the user. The frequency of sampling and the size of each sample is determined empirically according to the needs of the experiment.

ROTAMETER
A variable area flow meter which indicates the rate of gas flow into a fermentor. A manual valve is adjusted until an indicator ball rises up a tube of increasing width until the required flow rate value is reached on a calibrated scale marked on the glass wall of the tube. The bottom of the ball should rest on the calibration line.

2.5 Alternative vessel designs

There are alternatives to using a conventional, 'stirred tank' fermentor. Usually, alternative vessel designs are tried when the standard vessel configuration does not allow adequate growth of the organism (e.g. animal cells are often disrupted by shear forces in fermentors with turbine impellors) or the scale-up criteria require a different design of bioreactor (e.g. production of large quantities of single-cell protein is cheaper on a large scale if air lift fermentors are used to eliminate energy costs associated with a drive system). A number of these special designs are available at the bench/pilot scale of operation to allow small-scale research into the suitability of a particular method.

2.5.1 Air lift

This vessel design eliminates the need for a stirrer system. A tall, thin vessel is the best shape with an aspect ratio of around 10 : 1 (height to base diameter). Sometimes a 'conical' section is used in the top part of the vessel to give the widest possible area for gas exchange. Sensors can be mounted in a steel base section, from a collar at the base of the conical section or from the vessel top plate.

ASPECT RATIO
The ratio of the height of a fermentation vessel to its diameter. Typically, vessels for microbial work have an aspect ratio of 2.5–3 : 1, while vessels for animal cell culture tend to have an aspect ratio closer to 1 : 1.

The culture fluid is both mixed and aerated by a stream of air which enters near the base of the vessel. A hollow pipe or draft tube in the centre of the vessel provides a 'riser' for the air (which is full of bubbles) to move upwards to the top of the vessel. If a very large vessel is used, the hydrostatic head of the fluid provides a pressurizing effect to the lowest region of the culture where the air enters and so increases the dissolved oxygen concentration. The draft tube is usually double-walled to allow heating and cooling using a thermocirculator system.

When the aerated culture fluid reaches the top of the draft tube, it 'spills over' and begins to fall towards the bottom of the vessel via the space between the outer wall of the draft tube and the inner wall of the vessel. A large head space above the top of the draft tube allows for easy gas transfer from the liquid to the gas phase which causes the density/specific gravity of the liquid to increase and so it descends to the bottom of the vessel. The descending liquid returns to the base of the vessel where it is re-gassed and begins to rise again (see Figure 2.2).

A common use of air lift fermentors is the growth of shear-sensitive cells such as plant and animal cultures. Also, the design has been used for the production of large amounts of biomass as single-cell protein.

2.5.2 Fluidized bed

The microbes/cells are trapped in a physical medium (e.g. alginate beads) and held in the vessel by a mesh. Medium is recirculated via a pump and this can be easily adapted to give a continuous/semi-continuous flow to allow the trapped cells to effect chemical changes on constituents of the medium without being washed out along with spent medium. This system is well suited to growth

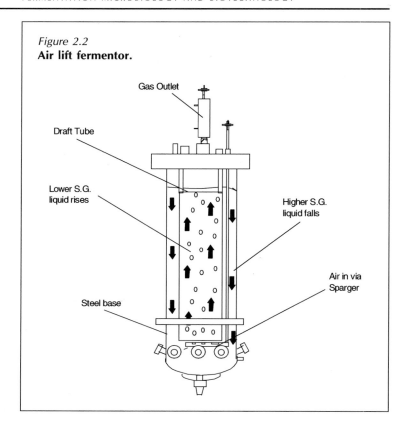

Figure 2.2
Air lift fermentor.

of animal cells on the smaller scale and has large-scale application in effluent/decontamination treatment plants.

2.5.3   *Hollow fibre*

This is a similar idea but the cells are now embedded in fibres contained in a cartridge which is bathed in circulating culture medium. This is often used for mammalian cell culture, where anchorage-dependent cell lines can be perfused with oxygenated medium. An extension of this method is to use cartridges containing two different bundles of fibres as a separation 'membrane' between a pair of fermentor vessels and allow dissolved gases and metabolites to be exchanged without cells crossing the barrier to allow, for example, toxicity or interactive studies.

2.5.4   **In situ** *sterilizable fermentors*

The use of autoclavable vessels of greater than 5–10 l working volume quickly becomes impractical. For safety and insurance considerations, vessels above this size are usually made of stainless

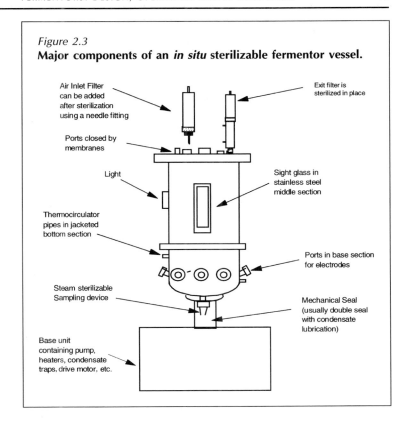

Figure 2.3
Major components of an *in situ* sterilizable fermentor vessel.

Air Inlet Filter can be added after sterilization using a needle fitting

Exit filter is sterilized in place

Ports closed by membranes

Light

Sight glass in stainless steel middle section

Thermocirculator pipes in jacketed bottom section

Ports in base section for electrodes

Steam sterilizable Sampling device

Mechanical Seal (usually double seal with condensate lubrication)

Base unit containing pump, heaters, condensate traps, drive motor, etc.

steel (316L) and are designed to be sterilized *in situ* using house steam or an electrical steam generator (which may be built into the base unit). The heating for the vessel is normally provided via a double jacket which can either be the full length of the vessel or cover just the bottom third. The bottom section contains large (25 mm) port fittings for electrodes and usually has some kind of steam-sterilizable sampling device/ harvest valve. The mechanical seal and drive shaft usually enter from the bottom on this size of vessel. As the vessel body is steel, a sight window and a light have to be fitted in order to see the culture. The vessel top plate has port fittings which use a membrane seal and port closure. Unlike the bench units, *anything* which dips into the culture liquid or is used as an addition port must be autoclaved separately and pushed through a membrane using aseptic techniques after the vessel has been autoclaved. In some fermentors, this even includes the air inlet filter and connecting pipe to the air sparger. The steam for sterilization is either supplied from an in-house source and used to heat the vessel jacket or is raised electrically from a separate steam generator. The medium in the vessel is heated to 121°C and often supplies the steam for sterilization of the exit gas filter (see Figure 2.3).

CONDENSATE TRAP
During *in situ*
sterilization of a
fermentor, spent steam
is condensed in the
condensate trap and the
resulting hot water
removed.

2.5.5 *Containment*

An *in situ* sterilizable fermentor may have to be altered in certain ways if the organism to be cultured is pathogenic or genetically modified. The alterations are designed to contain any release of microbes into the environment by using features such as a double mechanical seal, magnetic coupling, additional air filters, extra foam control systems and special sampling devices. More elaborate precautions include steam traces to all vessel fittings and discharge of any released liquid directly to a tank of disinfectant.

There are several different categories of containment (1–3 for pathogens and 1–4 for genetically modified organisms). Requirements differ at each stage and relate to building and working protocols as much as the design of the fermentor.

Applications for this sort of technology are medical research and vaccine manufacture.

2.6 Different types of instrumentation

Fermentor instrumentation can range from simple analogue control modules arranged as a 'stack', to powerful, embedded microprocessors which operate as on-board computers to directly operate the heaters, pumps, valves and other control actuators of a fermentor.

2.6.1 *Analogue controllers – rack system*

Here the individual modules for the measurement and control of each parameter are made to plug into a rack system for compact, flexible instrumentation. A bus system in the racking transmits signals from one module to another to allow measured values to be passed to controllers and for one controller to influence another, e.g. oxygen levels altering speed.

**ANALOGUE
INSTRUMENTS**
Measurement or control
modules which do not
contain processors. All
actions are 'hard-wired'
as electronic components
and any adjustment
is made using
potentiometers.

2.6.2 *Analogue controllers – separate modules in housings*

Each measurement/control module has a separate housing and essentially operates in isolation. The modules may be stacked one on top of another and can be supplied with 'pass through' power connections so that only one power socket is needed. This system allows for gradual building up or swapping of instrument modules between several fermentors.

2.6.3 *Digital controllers – embedded microprocessor*

The measurement sensors link directly to the single control module and several parameters are displayed immediately on a single screen. Control is by direct action on heaters, valves, etc. (Direct Digital Control or DDC). The microprocessor is permanently embedded in the instrumentation and may even be a single chip. Operation is usually via a simple menu system.

2.6.4 *Digital controllers – process controllers*

DIGITAL CONTROLLER
A digital controller uses a processor to store information about control output characteristics as mathematical algorithms. Consequently, changing the characteristics of such a controller is achieved by reprogramming the processor.

A complete process controller (usually from a production control environment) is added to a housing containing all the signal processing and control actuators for the fermentors. It exerts control in the same way as an embedded controller but is essentially a 'plug-in' component. The controller is usually 'programmed' using simple commands to input set-point values, etc.

2.6.5 *Digital controllers – direct computer control*

In this case, there is no external instrumentation or processor between the actuators and the computer. A printed circuit board with operational amplifiers for the probe input signals is the only electronic part which may be present in the fermentor base unit. A special Input/Output (I/O) card is needed for the computer which allows the input values to be accessed by the measurement and control software and sends signals out to operate relays, e.g. to turn a heater on or off. The processing power and speed of modern PCs make them more than capable of replacing separate PLCs or embedded controllers.

The advantage with this system is that the computer display and control software is integrated totally with the fermentor. Fewer components means that these systems are usually less expensive. However, the fermentor cannot be used without the computer and there is no backup for the control systems should the computer develop a fault and 'hang'.

2.7 Common measurement and control systems

TACHOMETER
An electronic device usually integrated into a drive motor to provide feedback about rotational speed in the form of an analogue signal.

2.7.1 *Speed control*

Speed control relies on the feedback from a tachometer located within the drive motor determining the power delivered by the speed controller to maintain the speed set-point value set by the user. A digital display shows the actual speed in rpm, as determined by the tachometer signal. A power meter is sometimes included which indicates how hard the motor has to work to maintain the set speed and thereby, indirectly, the viscosity or 'density' of the culture fluid. A DC, low voltage (24–50 V)

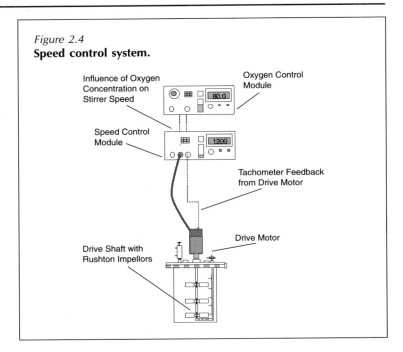

Figure 2.4
Speed control system.

Influence of Oxygen Concentration on Stirrer Speed

Oxygen Control Module

Speed Control Module

Tachometer Feedback from Drive Motor

Drive Motor

Drive Shaft with Rushton Impellors

motor is often used for safety reasons. Speed range is typically from 50 to 1500 rpm (exceptionally up to 2000 rpm).

Where speed is used to control the level of dissolved oxygen, an external signal from the oxygen controller can have an effect on stirrer speed. In this case, an absolute maximum and minimum value for speed can be set on the speed control module to limit the effects of the oxygen controller (which could set either too low a speed and impair mixing or too high a speed and cause excessive foaming) (see Figure 2.4).

2.7.2 *Temperature control*

A thermocirculation system around a vessel jacket has been chosen as an example here because it is the most complex of all the methods of temperature control. For simple direct systems such as a heater pad, it is simply a matter of fixing the heater, setting the desired temperature and switching on. Cooling is normally via a cold finger and flow of cooling water is controlled via the action of a solenoid valve. The Pt-100 sensor provides the feedback signal which causes the controller to take one of the following actions:

- heat at full power as the actual temperature is some way below set point;
- pulse the heater power as the actual temperature is close to set point;
- turn on the cooling valve as the actual temperature is above set point.

Indicator lights usually show which action the controller is taking at any given moment. A low-voltage DC heater system, e.g. 50 W, is often chosen for safety reasons. A circulation pump and pipework are added to the system for water circulation and any heating is indirect, i.e. on the water circulating in the vessel jacket. In this case, the connections to cold tap water must be made (securely, using jubilee clips or cable ties) and a drain pipe provided from the overflow point to a sink with a clear fall to the drain, i.e. the sink must be the lowest point for the whole length of this pipe. The water should be delivered from the mains at a minimum pressure of 2 bar and a flow rate of greater than 5 l min^{-1}. The water hardness should be no more than 50 ppm suspended solids to protect the heating elements from 'furring'. The vessel must be connected to the circulation loop (normally by rapid coupling connectors and flexible pressure tubing). Water is first supplied by opening a manual valve until the jacket is filled. The manual valve is then used to maintain just a trickle of water going through the jacket to the overflow to compensate for evaporation losses, e.g. 30–50 drops per minute (see Figure 2.5). The heating and cooling is controlled in exactly the same way as a directly heated system but only the water in the jacket

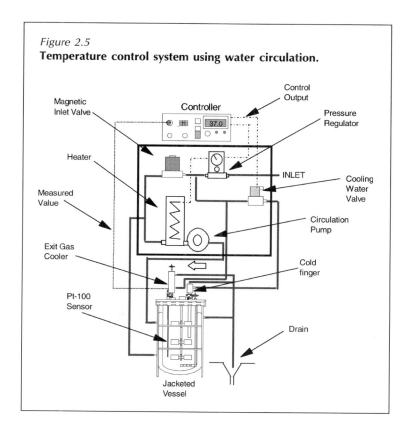

Figure 2.5
Temperature control system using water circulation.

is affected. The jacket provides a large surface area in contact with the vessel wall for heat exchange. Good temperature control can be achieved from approximately 5–8°C above the ambient temperature or 5°C above the temperature of the cooling water. Counter-cooling with water ensures stable temperature control when operating near ambient temperatures. Measured range is typically from 0 to 60°C (exceptionally up to 90°C).

2.7.3   *Control of gas supply*

The supply of gas (normally air) to the fermentor vessel is provided from a compressed air supply (oil-free) which may provide air for more than one vessel depending upon its size and output. A pressure regulation valve ensures that air reaches the vessel at a maximum of 1.5 bar.

The rotameter controls the actual flow rate of air through the fermentor. This should not exceed 1.5 vessel volumes per minute or droplets of water may become entrained in the stream of gas leaving the fermentor and so wet the exit gas filter, causing it to block. A valve at the bottom of the rotameter is turned and the indicator 'ball' in the rotameter tube rises or falls in proportion to the valve position. A scale on the tube gives flow rates in ml min^{-1} or l h^{-1}.

The air passes through the inlet air filter (Figure 2.6) which prevents any microbes from entering the vessel via this path. The end of the sparger is typically a ring with small holes through which the air is forced. The bubbles are immediately broken up and dispersed by the impellors on the drive shaft and the baffles which can be fitted near the wall of the vessel. The use of several impellors ensures all regions of the vessel receive good aeration.

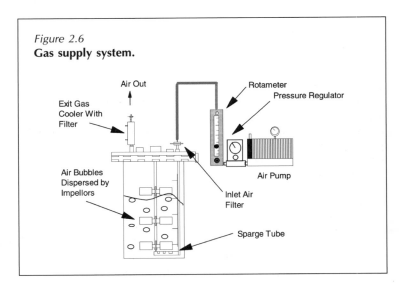

Figure 2.6
Gas supply system.

A 'head space' of around 20% is normally left between the culture level and the vessel top plate. Sometimes, gas can also be introduced into this region via a short pipe in the fermentor top plate. For certain types of fermentation, e.g. mammalian cell culture, a gas mixing station can be used to pre-mix several gases before they are introduced into the fermentor.

GAS MIXING STATION
A device used for animal cell culture which allows a mixture of air, oxygen, nitrogen and carbon dioxide gases to be blended into any desired combination before they are introduced into a fermentor. This allows great flexibility in how dissolved oxygen concentration and pH are controlled within the culture.

2.7.4   Control of pH

pH control is achieved by the addition of either acid or alkali to correct changes in the pH of the culture during growth. The controller uses a pH electrode (either glass or gel type) to sense these pH changes and provide a feedback signal. The pH measurement system is identical to a bench pH meter and is calibrated in the same way *before* the electrode is autoclaved. Temperature compensation during operation may have to be set manually or may be automatic. The calibration normally requires that the temperature is manually set to equal the temperature of the buffer solutions. Autoclavable electrodes have a limited life (20–50 sterilizations typically). The pumps used to supply the acid and alkali are normally built into the control module or the base part of the fermentor. Measured range is typically pH 2–12.

The reagent bottles are connected to the fermentor via transfer lines of silicone tubing. The bore of tubing selected will determine the volume of acid or base added by each activation of the relevant pump by the controller. Selecting the concentration of the acid or alkali will determine how much effect each dose has on the vessel contents. Normally, solutions of between 0.2 and 0.5 M of acid and base are used as a starting point. Using an ammonium salt for alkali addition can provide an extra nitrogen source for the growing culture.

A set-point value is set on the controller and/or an upper and lower limit to provide a 'dead band' range in which the controller is not active. This band should normally be around 0.5 pH units each side of the desired value to prevent overdosing of acid and alkali sequentially. A proportional band adjustment may be present to widen or tighten the range of pH value over which the controller acts and even a switch to give a non-linear control if good control cannot be achieved by any normal adjustments (see Figure 2.7).

DEAD BAND
An area around a set-point value which can be set where no control action will take place even if the actual value deviates from the set point. This is used especially for a parameter such as pH where small changes are often not critical and attempts to control them would lead to excessive controller action.

2.7.5   Control of dissolved oxygen

Dissolved oxygen is one of the most important and yet one of the most difficult parameters to control properly during a fermentation. The electrodes used to measure dissolved oxygen are of two distinct types:

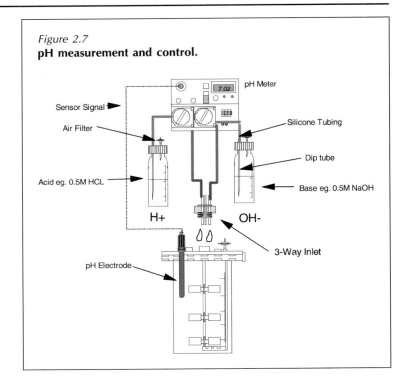

Figure 2.7
pH measurement and control.

1. Galvanic electrodes, which are generally simple, cheap, slow and have a limited life.
2. Polarographic electrodes, which are generally more complex, accurate, rapid and robust but expensive.

The polarographic electrode is used more commonly and will be the example provided. The key point with this type of electrode is that it requires a voltage to polarize the anode and cathode of the detecting cell. This polarization can take between 2 and 6 hours to complete (in an emergency, 40 min is just sufficient for most tasks). During this time, the electrode must be connected to its relevant module which in turn must be switched on.

The membrane of the polarographic electrode needs periodic replacement and a special cartridge kit is available from the manufacturer to make this a simple task.

The electrode can be 'zeroed' prior to autoclaving using a special zeroing gel provided by the manufacturer. The electrode is simply inserted into a sachet of gel, left for a short time and then the displayed value adjusted until it reads zero. Alternatively, the electrode can be zeroed after autoclaving by passing nitrogen gas through the vessel for some minutes and then adjusting the zero point.

The 100% value is a relative setting made after autoclaving and polarization of the electrode. The air flow and stirrer speed are turned to the maximum needed for the fermentation and the vessel left for some minutes. The display is now adjusted until it reads 100%.

Control of dissolved oxygen can be purely by speed control, purely by control of a proportional air valve, or by a combination of both. Speed adjustment may interfere with good mixing or increase the degree of foaming. Adjusting the air flow rate may also affect levels of foam. The most accurate form of flow control is to use a thermal mass flow control valve which measures and controls air flow based on the cooling effect the gas exerts when passed over a heated element. A simple option is the use of a solenoid valve with a manually set 'bleed' valve to simply vary air supply between a maximum and a minimum flow rate (see Figure 2.8). Range is typically 0–120%.

2.7.6 Antifoam control

The control of foam is based on its detection by a conductance probe in the vessel head space, which leads to the controller delivering a dose of liquid antifoam reagent via a peristaltic pump. A delay timer ensures the antifoam reagent has adequate time to reduce the foam level before another 'shot' of antifoam is added. The sensitivity of the probe to foam can be adjusted by the user to suit the conditions prevailing in the fermentor. A sheath of inert material around the probe prevents splashes

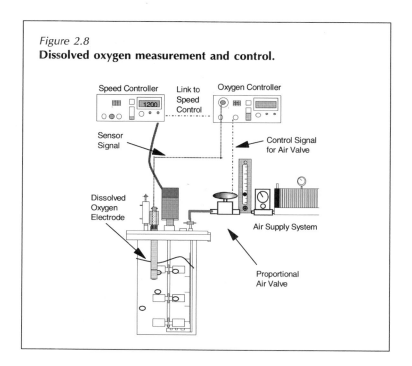

Figure 2.8
Dissolved oxygen measurement and control.

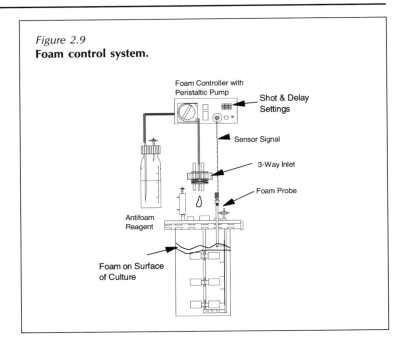

Figure 2.9
Foam control system.

of foam thrown up from the culture from giving 'false positives' (see Figure 2.9). Normally, the metal top plate is used to provide the electrical circuit for the probe to operate so a flying lead is provided which fits into a socket somewhere on the top plate.

Antifoam reagents can be mineral oils, vegetable oils or certain alcohols. Commercial preparations are available for use with cultures which ultimately provide materials for pharmaceutical manufacture. The key thing with using oils is that they can form a 'skin' on the surface of the culture and interfere with gas transfer at the liquid/air interface. If foam is allowed to build up unchecked it can eventually reach the exit gas filter so blocking it and providing a path for contamination.

2.8   Additional sensors

There are several additional process values which can be measured and controlled within a fermentor. The most common are listed below:

2.8.1   *Redox*

This refers to the reduction/oxidation (redox) potential of a system and can be applied to biological cultures as well as chemical reactions. Redox values are given in millivolts and usually refer to the oxidation or reduction potential of a given chemical relative to a standard hydrogen half-cell (which is defined as having a redox potential of 0.00 V). A half-cell which is more reducing than the hydrogen standard has a negative value and a stronger oxidizing agent will have a positive value. Reducing agents donate an electron to oxidizing agents. Depending on conditions (the pH can affect redox potential), this reaction is reversible.

The complexity of a typical fermentation broth (in terms of the sheer number of separate reactions going on at any one moment) means that the actual measured value is only of limited importance. However, for anaerobic cultures, the redox electrode provides a very sensitive 'watchdog' for the presence of oxygen. Tiny amounts of oxygen will have a relatively large effect on the redox value. A redox value below −200 mV is a good indicator of anaerobic conditions.

The redox electrode is very similar to a pH electrode and the conditions for handling and care are almost identical. An electrical zero can usually be set and shorting out the electrode connections should give a reading close to 0 mV. There is no calibration procedure as such. A check on the electrode can be made using ordinary pH buffers which have been saturated with an excess of solid quinhydrone. The quinhydrone dissociates into equal quantities of benzoquinone and benzoquinol below pH 8.0. The redox potential for this system at pH 7 should be around +285 mV and changes by 59 mV for each pH unit, e.g. pH 9 = 403 mV. Deviations of ±5 mV are acceptable. If an analogue meter is used, a variable offset is usually provided for output to a chart recorder.

2.8.2 Air flow

This has been mentioned previously in connection with dissolved oxygen control. A simple system would use a magnetic sensor and a variable area flow meter (rotameter) to measure the current air flow and adjust a motorized valve between 0 and 100% open or closed to control the flow. Alternatively, a solenoid valve could pulse the air flow very rapidly on and off between a maximum and minimum value set manually by the rotameter and a 'bleed valve' respectively. In this case, there is no displayed value for flow rate which could be used for data-logging or control by computer. A more sophisticated system uses a thermal mass flow control valve which combines both the measurement element and the control valve in one compact unit. In this case, measurement of flow is by detecting the change in temperature of a heating element before and after gas passes over it. The temperature difference is proportional to flow rate and this can be processed to give a measurement signal for logging and/or control. The practical points to consider when using this system are to ensure the valve is the right way round (an arrow on the metal section shows the direction of flow, however, this is sometimes obscured by the valve mounting) and that the control valve either bypasses the rotameter or the rotameter has a maximum value set to act as a physical limit on the amount of gas flow into the fermentor. Of course, the flow control systems can measure any gas flow, not just air. They can be used in combination under computer control to give a precise gas mix for any total flow rate using one mass control valve per gas and post-blending the mixture. However, this tends to be an expensive option. A simpler system uses solenoid valves to introduce each gas into a pre-mixing chamber at manually set flow rates and then controls the final gas flow into the fermentor with a single mass flow control valve.

2.8.3 *Weight*

This parameter is useful for continuous culture or fed-batch work where the rate of addition of feed must be known and controlled very accurately. It is possible to mount a whole fermentor on a balance and 'tare' out its weight. However, a more precise approach is to use a system with a small load cell mounted in such a way that only the vessel and its contents are measured. The 'deformation' of the load cell provides an output which can be transduced into an electrical signal. Control is normally by setting a maximum and/or minimum weight such that a feed pump operates whenever the value is less than the maximum. At the maximum value, a second pump operates to remove culture until the minimum weight is reached.

A variant of this system can be used to measure very accurately the amount of reagent, e.g. alkali, added to a fermentor. In this case, the reagent bottle is placed on an analytical balance which has a computer output (normally RS232) and the decrease in weight used as a measure of the amount of reagent added.

LOAD CELL
A method for measuring changes in weight in a fermentor vessel using deformation of a crystal as an indicator of load. Load cells are used where it would be impractical to use a conventional balance.

2.8.4 *Pressure*

The use of pressure as a control parameter is more common in larger, *in situ* sterilizable fermentors, for the obvious reasons that the vessel is made of steel and is already adapted to working at pressures of several bar. Also, it is typically in the larger vessel that maintaining the dissolved oxygen concentration at a high level presents a problem (an increased pressure increases the amount of oxygen which can be dissolved in the fermentation broth). However, even a small glass vessel can often operate at some overpressure (e.g. 1–1.5 bar) but it is vital that this is first confirmed by the manufacturer.

The measurement and control is the same whatever the size and type of vessel, except that the size of the control valve would alter. Measurement is normally provided by a piezo-electric sensor which can be mounted into a port closure and fitted to the vessel top plate. As with a load cell, it is the deformation of the crystal by the internal vessel pressure which gives a proportional electrical signal, typically in the range of 0–2 bar. The control element is a proportional valve which restricts the flow of gas out of the fermentor and thereby creates a 'back-pressure'. A mechanical overpressure valve or burst-disc is an essential safety requirement if overpressure is to be used.

PROPORTIONAL VALVE
A valve which can be adjusted electrically or pneumatically from 0 to 100% open or closed. The action of the valve is in proportion to the degree of change required by the controller to, for example, maintain a certain level of dissolved oxygen by adjusting the air flow rate.

2.8.5 *On-line measurement of biomass*

A direct measurement of the number of organisms in a culture is clearly desirable for the control of feed rate, oxygenation and general process optimization. Normally, measurements of total cell numbers, wet weight, dry weight and viable cell counts would be made by taking samples of the culture at set times and performing the relevant laboratory analysis. The results would be entered some minutes, hours or even days later, when they would be of little value for real-time control. Fluorescence microscopy has also been used as a measure of viable cell density either with a probe or externally through the vessel glass. Two probe-based systems which are reasonably accurate and can be used with bench-scale vessels will be described by way of examples below.

2.8.5.1 *Mettler–Toledo FCS turbidity system*

This is actually a nephelometer as it measures the scattering of light by particles in the culture (i.e. microbes). The advantage over a standard nephelometer is that the sensor is directly mounted in the vessel and connected to the electronics via a fibre-optic cable. A two-point calibration procedure is used and results can be expressed in a variety of units. One of the restrictions of such a system is that if the culture is very thin, then shiny, reflective components close to the sensor could affect the readings. Conversely, too thick a culture would not allow the light scattering to be detected properly. However, a range up to 250 g l^{-1} dry weight is measurable. The probe is resistant to fouling over time.

2.8.5.2 *Capacitance/conductance-based biomass monitor (Aber Instruments)*

This uses a totally novel measurement system to give viable cell numbers for all types of organisms (bacteria, yeast, filamentous fungi and animal cells). A probe in the vessel uses a radio-frequency electrical field to measure the natural capacitance of living cells with an intact plasma membrane. This build-up of charge is proportional to the cell concentration and specific to different types of cell. The signal is processed and can be expressed as dry weight or concentration in cells ml^{-1}. An anti-fouling system is used to maintain the ability of the probe to give accurate readings over long time periods.

This type of instrument makes possible the control of feed pumps by biomass concentration, and could form the basis for an automated transfer system for delivering a seed culture to a larger vessel after an inoculum of sufficient concentration has been produced in a smaller fermentor.

2.9   Simple continuous culture

There are a number of different ways of setting up a continuous culture. The method of choice has the virtue of being able to be added to any fermentor vessel without modification or the addition of extra controllers beyond the acquisition of a feed pump and a harvest pump. The feed pump is variable speed but the harvest pump can be fixed speed as long as its speed is greater than that of the feed pump. The harvest pipe, being sited at the surface of the culture, has the disadvantage of taking off only the top of the culture (which may not be representative of the bulk culture) but the advantage of helping to prevent build-up of foam. If foam is a problem then antifoam reagent can be added to the medium feed at a background level, e.g. 1 part in 20 000. Samples can be taken by putting a 'Y' piece in the take-off tubing and placing a sampling device in one arm of the 'Y' (see Figure 2.10).

2.10   Additional accessories and peripherals

2.10.1   Feed pumps

A separate feed pump is necessary for fed-batch or continuous applications. The type of pump used will depend on the size of the vessel, the required flow rates, the nature of the additive (e.g. viscous, corrosive, etc.) and the speed at which the addition should be made. Often, a feed pump is included as part of the basic fermentor or provision is made for its inclusion as an optional accessory.

2.10.1.1   Peristaltic pump

This is the most common type of addition pump used in fermentor applications and is often found in a fixed-speed version

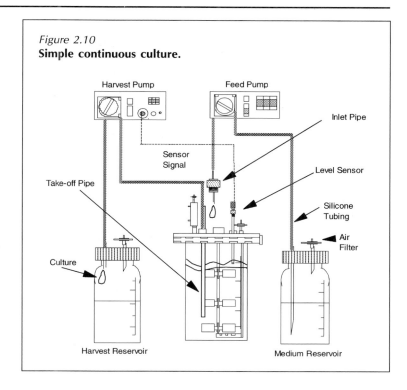

Figure 2.10
Simple continuous culture.

for addition of reagents such as acid and alkali. The pump uses rollers to occlude sections of silicone tubing sequentially. This action creates a 'peristaltic' flow of liquid through the tubing (analogous to the movement of food through the body by waves of peristalsis). The more rollers a peristaltic pump has, the smoother the flow. The pumps used for medium feed are typically variable speed units which can accept signals from an external source such as a computer or process controller to allow adjustment of feed rates based on, for example, respiratory quotient (RQ) or optical density (OD). The key advantages of peristaltic pumps are as follows:

- The liquid being pumped is not contaminated by any part of the pump.
- The pump is not contaminated by the liquid being pumped.
- A wide range of flow rates is possible using tubing of different bores.
- The accuracy of delivery is good over a wide range of flow rates.
- The pump can run safely even when the tubing is empty.

The disadvantages are:

- The flow is pulsed and cannot 'push' against high back-pressure.

RESPIRATORY QUOTIENT (RQ)
RQ is a mathematically derived value related to the use of oxygen by a microbial culture as compared with the evolution of carbon dioxide. This value can be used to adjust feed rates of sugars to manipulate microbial physiology.

- The tubing wears over time and flattens, reducing accuracy and increasing the risk of bursting.
- Silicone tubing is not suitable for all liquids and alternative tubing material may be too rigid to allow correct peristaltic action.

The problem with tubing wear can be partly overcome by leaving extra length in the tubing when connecting it to the fermentor vessel and moving a new length into the pump head every few days to stop wear in just one small section.

2.10.1.2   Syringe pump

This type of pump works on a completely different principle and is useful for fed-batch applications rather than continuous medium feed. In this case, a syringe (of various sizes from μl to 100 ml, typically) is filled with the liquid to be added and secured onto the syringe pump. This is done in such a way that the syringe plunger is linked to a movable piston and the syringe barrel is held firmly to the main body of the pump. A flow rate is set (in mm min^{-1} or directly in ml min^{-1} or ml h^{-1} if the pump can be calibrated). The piston is then moved at the desired speed by the action of toothed gears linked to a stepper motor engaging with 'teeth' on the underside of the piston. When the syringe is empty, an alarm condition is indicated and the pump automatically stops or, on very sophisticated pumps, a signal can be sent which allows automatic refilling from a reservoir. This type of pump is typically used in medical applications for the precise delivery of drugs. Most now allow for computer control.

The advantages of syringe pumps are:

- Very precise and accurate delivery; non-pulsile flow.
- Very low flow rates possible.
- Can work against back-pressure.
- Can pump very viscous fluids, e.g. oils.

The disadvantages are:

- Must be replaced or refilled regularly.
- Range of flow rates not as wide as with peristaltic pumps.
- Usually more expensive.

2.10.1.3   Diaphragm/positive displacement pump

This type of pump is a hybrid of the two previous types in that it uses a flexible membrane and soft tubing to displace the liquid and control its flow but also employs a piston and pump heads of fixed volume. A stainless steel pump head is connected to soft tubing and autoclaved. It is then located onto the pump and held securely. A central piston and two small 'pinch valve' pistons move in and out to push liquid out of the central chamber of the pump head and allow fresh liquid to refill the chamber on each half of the pump stroke. The flow rate is controlled by the frequency of the pump stroke and the size of the pump chamber (which can usually be exchanged). A return spring returns the diaphragm to the correct position after each inward stroke.

A variant of this is to use a 'captive syringe' where the piston moves in and out of the chamber, emptying it fully then drawing back to refill on each stroke.

Advantages of this type of pump include:

- Accurate delivery.
- More choice in type of tubing for corrosive liquids, for example.

Disadvantages are:

- Less flexibility on flow rate ranges for each pump head.
- Stainless steel chamber and diaphragm may not be suitable for some liquids.
- Pulsed flow even more pronounced than with a peristaltic pump.

2.10.2 Exit gas analysis

Measurement of the amounts of different gases leaving a fermentor vessel can provide valuable information about the metabolic processes taking place under the conditions in which the culture is growing. For example, the ratio of oxygen and carbon dioxide entering and leaving the vessel, or respiratory quotient (RQ), can be used to determine whether a yeast is producing biomass or alcohol. The RQ can be calculated by most fermentation software packages and can then be used as part of a control algorithm so that the flow rate of a feed pump can be altered accordingly. The entry gas does not need to be analysed providing it is air, because the amounts of oxygen and carbon dioxide will be those of the atmosphere. The flow rate of the air into the vessel will need to be accurately measured – normally by using a thermal mass flow controller. The exit gas may need to be conditioned (e.g. moisture removed) before going into the analyser, depending on the type of instrument used.

2.10.2.1 Infra-red carbon dioxide analyser

A hot wire is used to generate a source of infra-red radiation which passes through the gas which comes from the fermentor into the sample chamber. An infra-red detector measures the amount of radiation reaching it after some is absorbed by the gas. An optical filter is used to make sure the detector only responds to the gas of interest. A 'chopper' or rotating shutter is used to allow the detector to see a reference source at regular intervals. This type of detector allows continuous measurement with good sensitivity, accuracy and selectivity. Output signals are normally provided either as analogue signals (e.g. 0–10 V) or a string of ASCII characters for printing or transfer to computer software.

2.10.2.2 *Paramagnetic oxygen analyser*

Oxygen has a particular physical property which this type of analyser can utilize, i.e. that it is more susceptible to a magnetic field than other gases. This property is measured using a finely balanced test apparatus suspended in the test chamber of the analyser. A dumb-bell of gas-filled spheres is linked to a support mechanism suspended in a magnetic field created by permanent magnets. If oxygen is present in the test gas, its attraction to the magnetic field will cause the dumb-bell to be displaced. The movement is detected using a mirror in the centre of the balance system which displaces a beam of light shone via the mirror onto a photocell. The signal from the photocell will be proportional to the amount of oxygen present in the sample. A low flow rate is needed for this system to work properly, so inlet gas is typically pumped into the detector cell with most being discarded through bypass pipework. Once again, suitable analogue and computer outputs are normally provided.

Both infra-red detectors and paramagnetic oxygen analysers need to undergo a calibration procedure before use.

2.10.2.3 *Mass spectrometer*

This represents a step upwards in versatility, capability and also cost! A wide range of gases can be analysed (both in a gas stream and as dissolved gases in the vessel), e.g. oxygen, carbon dioxide, argon, nitrogen, ammonia, hydrogen, methanol, ethanol and several other organic volatiles.

The analysis is rapid, so allowing a multi-inlet system to be used which 'shares' the analyser between separate vessels within a bank of fermentors. Mass spectrometers can be magnetic sector, multi-collector or quadropole. Each has its own particular advantages but the magnetic sector can be used to illustrate the principles upon which they all work. The physical principle involved is the ionization of gas molecules by an electron source, usually a hot filament, followed by their separation in a magnetic field (the quadropole analyser uses a combination of RF and DC electrical fields to do this job) according to their mass/charge ratio. This separation takes place in a virtual vacuum to

Box 2.1 **Calculation of RQ**

When data are obtained from an exit gas analysis, it is generally required that the parameter RQ is calculated in order to gain an insight into the metabolism of the organism under investigation. This is normally done by programming a definition into a computer software package for data-logging and control. Here is a generalized outline of the steps needed:

RQ = CPR (Carbon dioxide production rate)/ OUR (Oxygen uptake rate)

CPR = [Flow CO_2 out] × [Conc'n CO_2 out] − [Flow CO_2 in] × [Conc'n CO_2 in] gmol/O_2/h

OUR = [Flow O_2 in] × [Conc'n O_2 in] − [Flow O_2 out] × [Conc'n O_2 out] gmol/O_2/h

Concentrations of gases in inlet gas (AIR) O_2 = 20.95% CO_2 = 0.03%

The working volume of the fermentor (WV) and the gas flow rate (FR) must be known to make the calculation:

VUO_2 (Volumetric uptake O_2) = (FR × (0.2095 − (Exit O_2/100))) WV

$VPCO_2$ (Volumetric production CO_2) = (FR × ((Exit CO_2/100) − 0.0003))) WV

Typical values for Exit O_2 could reach approximately 16% and for Exit CO_2 around 5%.

$$RQ = \frac{VPCO_2}{VUO_2}$$

Values of RQ for yeast biomass fermentation, for instance, would need to be kept around 1 for optimal biomass production.

A worked example of the above definitions:

WV = 1 l, FR = 1 l min^{-1} (1 vessel volume of gas per minute, a typical flow rate).

Exit O_2 = 16% and Exit CO_2 = 5%

VUO_2 = 10/10 × (0.2095 − 0.16) = 0.0494

$VPCO_2$ = 10/10 × (0.05 − 0.0003) = 0.497

RQ = 0.497/0.494 = 1.006

To achieve control, a sequence along the lines of the following would have to be set up:

if RQ < 0.95 then feedrate = feedrate × 1.05

if RQ > 1.05 then feedrate = feedrate × 0.95

minimize collisions prior to sorting. The magnetic field is 'tuned' so that only the ions of interest will be focused onto the detector system – a Faraday Cup. This is a metal plate which generates a tiny electric current whenever a gas ion strikes it. This signal can be amplified and displayed. By re-tuning the magnetic field very quickly (milliseconds) different ions can be detected and measured in a sequential way which is so fast it appears to give an almost simultaneous measurement.

2.11   Fermentor preparation and use

2.11.1   *Disassembly of the vessel*

The fermentation is shut down from the control unit and transfer lines plus cable connections removed. Following a fermentation, the vessel should be re-autoclaved, taking care that inlets and outlets are properly prepared. The culture should be disposed of as laid down by departmental safety procedures The clip/clamps/bolts which retain the vessel top plate are undone until the whole assembly springs free. The top plate can now be lifted upwards away from the glass section of the vessel, taking care that the air sparger, drive shaft/impellors and Pt-100 temperature probe completely clear the vessel. The flanged top section of the glass vessel and seal can now be seen.

2.11.2   Cleaning

The pH and dissolved oxygen electrodes should be removed and stored in suitable reagents according to the manufacturer's instructions. Periodic cleaning and regeneration of the electrodes are also covered by these instructions. Doing this maintenance is very cost-effective. The vessel should be rinsed several times in distilled water to remove any loose culture residues. Cleaning of growths of culture on the vessel walls may require disassembly and light brushing of the glass. At this point, an examination of any chips or cracks in the vessel glass can be carried out and a replacement made if necessary. Vessels must be stored clean and dry. In use, any spillages of reagents or medium should be wiped up immediately with a damp cloth and not be allowed to dry out. Contact between the top plate and liquids with a high chloride ion content (e.g. common salt solutions, hydrochloric acid) should be avoided to prevent corrosion. The pump heads and covers must be thoroughly cleaned if a tube breaks and reagent leaks out.

2.11.3   Preparations for autoclaving

The vessel seal should be removed and lightly greased with a suitable silicone grease. On replacing the seal, it must be correctly located so that there is no chance of any part lifting or kinking. At this point, the vessel can be filled with medium to a maximum of 80% full (if active aeration is to be used this space is vital for gas exchange). The minimum medium volume is the amount needed to cover the electrodes adequately. The vessel top plate can now be replaced and any clamping ring or bolts tightened firmly. The ports for electrodes have O-ring seals which should be lightly greased with a silicone preparation before autoclaving. Electrodes normally push directly into the port fitting and the collar is tightened down to compress the O-ring seal. All other fittings such as pipes are fitted in the same way. Ports not in use have 'stoppers' fitted, and their O-ring seals should have a light coating of silicone grease applied. The pH electrode should be calibrated in appropriate buffers i.e. pH 7, then either pH 4

Figure 2.11
Vessel and accessories prepared for autoclaving.

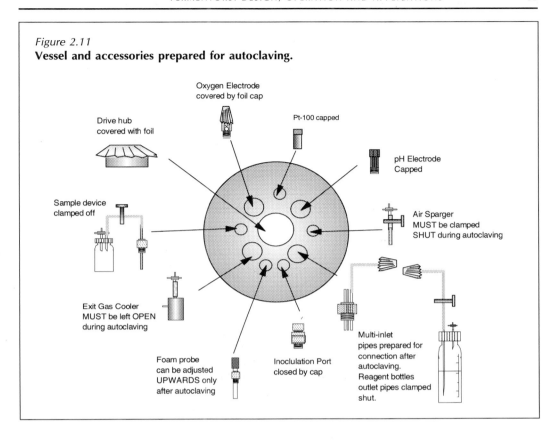

Oxygen Electrode
covered by foil cap

Pt-100 capped

Drive hub
covered with foil

pH Electrode
Capped

Sample device
clamped off

Air Sparger
MUST be clamped
SHUT during autoclaving

Exit Gas Cooler
MUST be left OPEN
during autoclaving

Foam probe
can be adjusted
UPWARDS only
after autoclaving

Inoculation Port
closed by cap

Multi-inlet
pipes prepared for
connection after
autoclaving.
Reagent bottles
outlet pipes clamped
shut.

or pH 9 (or just 4 and 9 for a 'high reference–low reference' calibration of a DDC controller channel). The pH and dissolved oxygen electrodes should be fitted, taking care not to damage them by careless insertion into the port. Both must be tightly capped to prevent moisture getting into the electrical contacts. For the dissolved oxygen electrode, a cap may have to be improvized from aluminium foil. The Pt-100 temperature sensor must be fitted and capped. If used, the foam probe is fitted so that it is above the liquid level in the head space (Figure 2.11).

Reagent bottles are prepared in a similar way to the fermentor vessel. A membrane is placed over the mouth of the filled bottle and an open-topped cap screwed into place. A short needle is connected to a disposable filter which is then used to pierce the membrane in the bottle. The needle must not dip into the liquid and nothing must block the free passage of air through the filter. A long piercing needle/pipe is used to puncture the membrane and push it into the liquid as far as possible. This pipe should now have a length of silicone tubing fitted to it which is long enough to reach the peristaltic pump heads of the fermentor instrumentation for which it is intended. The tubing is clamped

Box 2.2 **Preparation of tubing for aseptic connection**

Any silicone tubing to be joined needs the two ends to be prepared as follows:

1. The open (female) end of the tubing is simply covered in aluminium foil held in place with autoclave tape. For added security, the open end can be sealed with a short piece of glass rod flattened at one end to make a 'bung'.

2. The other (male) end has a short length of stainless steel pipe pushed into the silicone tubing. The exposed length of pipe is covered with foil which is taped closed.

3. Silicone tubing is connected to any of the inlet pipes intended for use for reagent addition. The tubing must stretch to the peristaltic pump heads when the vessel is in place.

4. The tubing is closed with clamps or any other closure system which will withstand autoclaving (the arterial clamps used in medicine are often employed where a connection must be opened and closed many times, e.g. to a sampling device). The tubing which goes to the sampling device from the top plate must be clamped shut during autoclaving.

so that no liquid can escape during autoclaving. A similar procedure is used for sampling and/or harvest bottles except that two short pipes are used so neither dips into the collected culture. Some preparation is needed at the autoclaving stage for the aseptic coupling of silicone tubing ready for operation.

The exit gas cooler should be fitted to one of the wider ports. A short length of silicone tubing should be attached to the top of the air outlet and a small 0.22 μm filter fitted. The air outlet line must be kept open during autoclaving. A short length of silicone tubing must be fitted to the air sparger inlet pipe with a 0.2 μm disposable filter mounted on top. The tubing between the sparger pipe and the filter must be clamped shut during autoclaving. If a port is to be used for inoculation or piercing with a needle, a silicone membrane must be fitted into the empty port and a clamping collar/cap used to hold it in place.

2.11.4 *Autoclaving*

The vessel and any reagent/sampling bottles already connected by silicone tubing are assembled together on a steel tray or in an autoclave basket. A final check should be made that at least one route is available for air to enter and leave the vessel(s) and that all lines dipping into liquid are clamped closed. If the vessel has top drive and a mechanical seal, the seal must be lubricated (normally with glycerine).

If the medium cannot be autoclaved, a suitable volume of distilled water should be used and removed via a sample line before the actual medium and inoculum are aseptically transferred. A quantity of liquid is certain to be lost during autoclaving (approximately 10%) so either the medium is over-diluted to compensate for this or sterile, distilled water is added afterwards to restore the volume. Some form of indicator such as autoclave tape should be included to provide a warning if the correct sterilization procedure has not been carried out. Autoclaving at 121°C for 30 min is normally considered adequate for vessel sterilization to minimize damage to the constituent chemicals of the medium. However, some work may require temperatures of 134°C for several hours to ensure sterility. If in doubt, your safety committee should be consulted. Also, the autoclave used must have good pressure equalization during the cooling down phase of operation to prevent medium being boiled off. The vessel and any accessory bottles must be allowed to cool completely before handling.

2.11.5 *Set-up following autoclaving*

Firstly, the silicone tubing prepared prior to autoclaving must be connected to the fermentor vessel (if the bottles they are connected to were autoclaved separately). Figure 2.12 shows how this is done.

The air sparger is connected to the rotameter by a piece of silicone tubing from the top of the filter to the air outlet of the rotameter. The air sparger line is unclipped between the metal pipe and the air filter. The exit gas cooler is connected to the

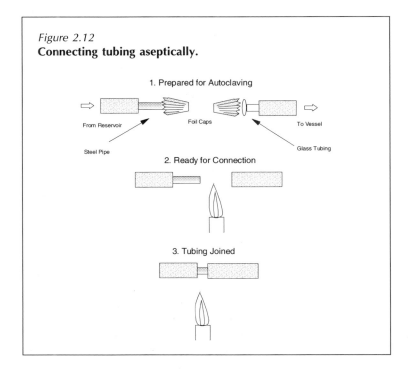

Figure 2.12
Connecting tubing aseptically.

water supply, either directly or via the fermentor base unit. The tubing for water in, out and drain is connected to the vessel jacket for a water system, and the water turned on so that the vessel jacket is filled. Alternatively, any pads or heater cartridges are connected to the base unit or temperature control module; the cold finger is connected to the water supply.

The tubing from the reagent bottles is connected to the multi-way inlet (if necessary), and the silicone tubing from the reagent bottles is located in the relevant peristaltic pump. Any aseptic connections must be made first if the reagent bottles were auto-claved separately from the vessel. The clamps are removed so liquid can flow freely. A manual switch is often fitted which allows the pumps to be primed with liquid prior to use. The drive motor is located onto the top plate (if appropriate), ensuring a good connection is made to the drive shaft. The Pt-100 temperature sensor is connected to the control module and the pH electrode by removing the shorting cap and screwing in the cable. The dissolved oxygen electrode is connected to the appropriate cable (this requires some care, but the connector should lock firmly when it is correctly positioned). Connections to the foam probe are made, usually one wire into the electrode and one on the vessel top plate to make a circuit. The mains electricity to the instrumentation modules is switched on, allowing some hours (minimum of 2, preferably 24) for the dissolved oxygen electrode to polarize properly. This delay is not needed for galvanic electrodes. Setting the temperature control at this stage will ensure the fermentor is ready to inoculate after calibration of the dissolved oxygen electrode. After polarization the dissolved oxygen electrode is calibrated and the air supply turned on. The maximum stirrer speed to be used and the maximum air flow required on the rotameter are set. After leaving for about 5 min, the 100% level is set. The fermentor is now ready to inoculate.

2.11.6 *Inoculation of a fermentor vessel*

This section assumes that all the set-up procedures listed above have been carried out. The simplest way to inoculate is to have a dedicated port fitted with a membrane which is capped off before autoclaving. The inoculum (which should normally be no more than 5–10% of the total culture volume) is aseptically transferred to a sterile, disposable syringe of a suitable size. The port fitting is removed and held vertically to prevent contamination of the bottom end. The syringe needle is quickly pushed through the membrane and the inoculum transferred into the vessel. The vessel may be actively aerated during this procedure to minimize the risk of a contaminant getting into the vessel (safety considerations permitting). The syringe needle is quickly

Figure 2.13
Inoculation of a fermentor vessel.

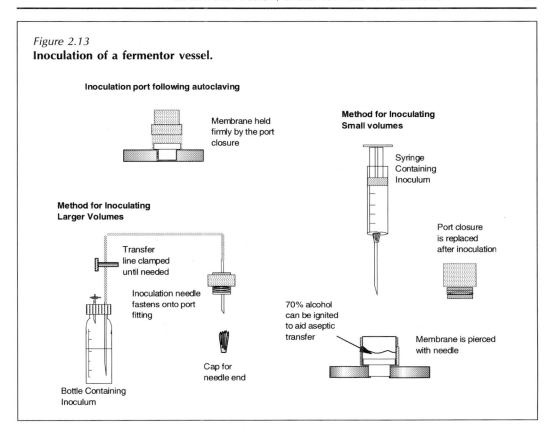

Inoculation port following autoclaving

Membrane held firmly by the port closure

Method for Inoculating Small volumes

Syringe Containing Inoculum

Port closure is replaced after inoculation

Method for Inoculating Larger Volumes

Transfer line clamped until needed

Inoculation needle fastens onto port fitting

70% alcohol can be ignited to aid aseptic transfer

Membrane is pierced with needle

Cap for needle end

Bottle Containing Inoculum

withdrawn and the silicone membrane reseals. The port fitting is now replaced. For added security, a couple of drops of 70% ethanol can be placed on the membrane surface before piercing and ignited to provide a thermal barrier. Alternatively, an 'aseptic connection' can be made to an inlet pipe. If the line connected has a 'Y' coupling in it, then the same aseptic connection could be used to introduce medium. This technique is useful if a dedicated inoculation port cannot be provided. Figure 2.13 shows the alternative methods.

2.11.7 Sampling from a fermentor vessel

All sampling starts with a sample pipe, which should dip into the bulk of the culture liquid. The simplest system is to have a length of silicone tubing connected to the sample pipe (and clamped off until a sample is to be taken) which is plugged and covered at the open end. After autoclaving, the tubing is quickly unwrapped, the plug removed and the open end dipped in a container of 70% ethanol. When a sample is to be taken, a sterile syringe is quickly coupled to the tubing and the sample with-drawn into the syringe. Aeration is stopped during this process to prevent 'surging' of culture into the syringe. Any culture in the transfer line is discarded into a 'kill jar' and the tube re-placed in the ethanol. Clearly, this method can be prone to con-tamination and is not suitable for organisms with any risk to operators. A better approach is to use a sampling device. At its simplest, this is a bottle which has two metal needles/pipes per-manently fixed through both the metal and rubber seals of its cap. One pipe has a 0.22 μm air filter connected and the second is linked by silicone tubing (clamped off until a sample is needed) to the sample pipe. A syringe is fitted to the air filter after auto-claving. The sample device is usually attached to the vessel top plate so that the glass bottle hangs down vertically beneath the cap and a supply of bottles of the same size are autoclaved ready for use. When a sample is to be taken, the clamp on the sample pipe tubing is released, and the syringe pulled back to create a partial vacuum in the sample bottle. Culture will flow into the sample bottle. When the required volume of sample has been taken (never more than 75% of the total bottle volume) the syringe is pushed in to clear the line and the sample tubing is clamped shut again. The sample bottle with the culture is quickly removed and capped with a sterile cap from a new glass bottle which replaces it under the sampling device. Figure 2.14 illus-trates the method.

It is often preferable simply to discard the first few millilitres of the next sample by using it as a 'wash' for the transfer line, putting a fresh bottle under the sampling device to actually take the sample.

2.11.8 Routine maintenance of fermentor components

2.11.8.1 pH electrodes

pH electrodes are a semi-consumable item. Typically, an elec-trode will need replacing after 20–50 autoclave cycles. This broad range reflects the different results sometimes found between different brands and types of electrode and the harshness of the environment in which they are used, e.g. high temperatures, frequency of autoclaving, etc. Ageing of electrodes can show up as a sluggish response time, and a smaller slope on calibration, especially in the alkaline region. The ageing process is speeded up by high temperatures, so an autoclavable electrode will in-variably have a shorter working life if repeatedly heated to steri-lization temperatures. However, during its working life, the pH electrode may need attention to ensure optimal performance. In this respect, the gel-filled pH electrodes are normally less demanding and are physically more robust than the glass, liquid-filled variety.

Figure 2.14
Sampling.

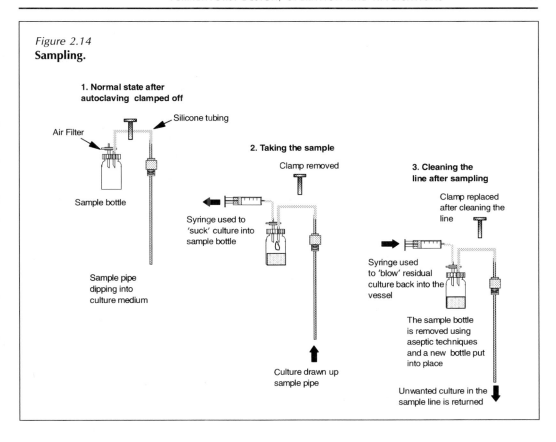

1. Normal state after autoclaving clamped off

Silicone tubing

Air Filter

Sample bottle

Sample pipe dipping into culture medium

2. Taking the sample

Clamp removed

Syringe used to 'suck' culture into sample bottle

Culture drawn up sample pipe

3. Cleaning the line after sampling

Clamp replaced after cleaning the line

Syringe used to 'blow' residual culture back into the vessel

The sample bottle is removed using aseptic techniques and a new bottle put into place

Unwanted culture in the sample line is returned

Box 2.3   Routine Maintenance of pH Electrodes

1. Always store an electrode which is not in use in liquid – preferably 3 M KCl solution, although pH 7 buffer or even fermentation medium will do (in descending order of preference).

2. An electrode which has completely dried out will need regenerating – a procedure which normally involves a brief immersion in a hydrofluoric acid solution (typically supplied by the manufacturer) followed by prolonged soaking in storage solution. This procedure may also regenerate 'aged' electrodes for a time. (Note that glass vessels cannot be used for this task due to the effects of the hydrofluoric acid.)

3. A build-up of protein material on the electrode membrane can be dealt with by soaking the electrode in a pepsin/ hydrochloric acid solution for about an hour.

4. If a discolouration (blackening) of the diaphragm has been caused by sulphides in the medium, soaking for several hours in an HCl/thiourea solution will recover its shiny surface.

5. Never wipe the tip of the electrode with a paper towel as this causes scratching.

6. Glass/liquid electrodes should have the reference electrolyte topped up periodically.

7. To test if the pH meter/cable are functioning correctly, a pH simulator can be used. If such a device is not at hand, a paper clip can be used to short the centre pin of the electrode plug and the outside metal ring of the cap. The pH reading should be very close to 7.0.

2.11.8.2 *Dissolved oxygen electrodes*

Galvanic electrode: This is a simple, self-powered system which relies on a lead anode and a silver cathode which are shorted together. Oxygen is reduced at the silver cathode and an electrical current flows in proportion to its partial pressure. The response time is slower than a polarographic electrode but the initial cost is far lower. The electrode is conditioned for the first time by the autoclaving process. The probe is relatively robust except if the lead anode is exposed to air due to a drop in the level of electrolyte in the probe. Therefore, all air bubbles must be removed when electrolyte is added and probes must always be stored upright with an adequate level of electrolyte and with the electrode shorted out (cap on). Periodically, the electrode membrane will need replacing. This involves soaking a section of new silicone rubber tubing in chloroform to swell it (safety hazard!). While this is happening, the old tubing and membrane is cut away. A new membrane is stretched tightly across the electrode tip and the silicone tubing placed over it to hold it in place without any wrinkling. The end away from the tip is trimmed off and secured with a little Araldite glue and the electrode tapped gently to remove air bubbles.

Polarographic electrode: These electrodes work on a different principle (the Clark Cell). The electrode consists of a silver anode/reference electrode with a platinum cathode encased in a glass rod. The end of the glass rod is shaped to accept a replaceable membrane cartridge. A temperature sensor is included for automatic compensation. The whole assembly is usually encased in a stainless steel housing. A chloride solution is used as an electrolyte. A polarization voltage must be applied to the electrode which can then reduce the oxygen passing through the membrane and generate a current. For this reason, the electrode must be autoclaved shorted out and be connected to a source of polarizing voltage for a minimum of 2 h before use. The membrane must be replaced periodically by unscrewing the bottom section of the electrode and gently removing the old cartridge. A new cartridge is fitted, having been filled with fresh electrolyte, and pushed onto the inner glass rod. If necessary, the small O-ring in the bottom section of the housing can be renewed before replacing the cartridge.

A functional check on the electrode can be performed by setting the reading of a polarized electrode in water saturated with air to 100%. The air is replaced with nitrogen gas and the value should fall to less than 1.5% after about 5 min. If the electrode is now removed into the air, the value should return to 98%+ within 45–90 s.

Box 2.4 **Replacing a mechanical seal**

The procedure for replacing a typical single mechanical seal is as follows:

1. Unscrew the top part of the drive shaft and remove. Measure the distance between the bottom of the retaining ring and the inner surface of the top plate. This allows the new seal to be correctly tensioned by setting the same gap at the end of the procedure.
2. Remove any retaining ring by loosening the small screw(s) holding it in place. Disassemble the retaining ring, hard seal, soft seal and bottom O-ring in order. Examine the faces of the static and rotating seals. Scoring of the surface indicates the seal has worn out. Never try to replace the old seal if it looks undamaged – it cannot be reused.
3. Insert a new O-ring into the 'hole' in the vessel plate which leads to the drive shaft bearings. This may be a tight fit but it must seat properly to prevent leaks. Replace the static seal by pushing it into the O-ring to obtain a snug fit. The protective rubber cover over the top face must be removed at this point.
4. Fit the rotating seal onto the drive shaft. Ensure the inner O-ring and metal washer are firmly pushed down into place as the tensioning spring will rest on these. Fit the new tensioning spring into place then push onto this firmly with the retaining ring until the spring is tightly compressed. Measure the gap with a ruler/calipers and only tighten the screw(s) on the retaining ring when the correct distance has been achieved.
5. Replace the drive shaft and gently turn the shaft by hand a small distance to ensure it moves freely. Lubricate with glycerine.

2.11.8.3 *Changing a mechanical seal*

Different arrangements of mechanical seal are used for different makes of fermentor. The type described is a good example as it shows many of the features used with the mechanical seals employed on both bench-scale and *in situ* sterilizable fermentors. Seals should be replaced regularly (at least every 6–12 months, depending on use).

2.12 Major types of organisms used in fermentation

A wide range of organisms can be used in fermentation for the production of various chemicals and antibiotics.

2.12.1 *Bacteria/yeast/fungi*

Either a natural isolate of the culture is made or an uncontaminated 'master' culture is revived from a 'freeze-dried' state or from liquid nitrogen storage. Bacteria, yeasts and filamentous fungi are the usual starting materials. The starting cultures are grown in petri dishes on agar containing the required nutrients and incubated at an optimum temperature, humidity and gaseous atmosphere. If necessary, shake-flask cultures are prepared in an incubator shaker to produce larger volumes of culture for analysis or for use as an inoculum for a fermentor. Flasks are typically filled to 20–25% to provide a large area for gas exchange and are 25 ml to 5 l in total capacity. An inoculum is prepared either from the shake-flask culture or by suspending colonies from growth on solid media in a nutrient/buffer solution. This is aseptically transferred to a fermentor vessel. The inoculum is typically 5–10% of the volume of medium in the fermentor and is often diluted in order to introduce a known number of cells/ml in this volume. A laboratory scale fermentor can be used to provide larger volumes (typically 1–5 l of culture). Growth is normally as a batch fermentation with the following parameters measured and controlled as necessary:

Temperature typically 20–40°C (4–90°C at the extremes)

Speed typically 150–1500 rpm (20–2000 rpm at the extremes)

pH typically 4–8 (2–12 at the extremes)

Dissolved oxygen typically 40–80% (0–100% at the extremes)

In some cases, a continuous culture system may be used to draw off culture of a known cell concentration for long periods. A concentration of a growth-limiting metabolite (chemostat) or an indirect estimate of cell numbers (e.g. turbidity, turbidostat) is used to keep the culture density constant for many days/weeks. Samples are taken from the fermentor for counting, culturing on solid medium or biochemical assay for a particular metabolite, sugar utilization, etc. On completion of the culture, the whole culture can be centrifuged, e.g. $4000 \times g$ for 15 min to separate whole cells from the culture supernatant fluid. Alternatively, a membrane or hollow-fibre system can be used for this initial separation.

The separated cells can be washed in buffer, resuspended, and can then be disrupted using a variety of methods, e.g. ultrasonic disruption, freeze–thaw, shaking with beads, etc. Supernatant fluids can be further separated and analysed using separation columns, fraction collectors and spectrophotometry. Gel electrophoresis, ELISA assay, etc. are also used. Applications are almost too numerous to mention but obvious categories include brewing (yeasts), antibiotic production (fungi), enzyme production (bacteria) and food manufacture such as yoghurt (bacteria). Environmental uses include removal of toxic wastes, biological control of crop pests and soil regeneration.

2.12.2 Plant cells

The whole plant is grown by conventional means. A sterile knife is used to cut a small section of root, stem or leaf which is transferred onto solid medium to produce a ball of cells (a thallus). The thallus is separated with a sterile blade and the cells grown in shake-flask culture in an incubator shaker at 20–30°C and 100–200 rpm. The cells are transferred to a fermentor under the following conditions: e.g.

• stirrer speed ≈ 100–150 rpm;
• temperature 20–30°C (cooling needed if lights used);
• pH not generally controlled as it tends to cycle naturally;
• dissolved oxygen rarely controlled as it is not generally a limiting factor;
• foaming can be a problem unless controlled.

Cells may be cycled between the shaker and fermentor several times to build up shear resistance. The cell growth is normally too slow for effective continuous culture. For material inside the plant cell a 'Potter' type homogenizer can be used for gentle disruption. Centrifugation at approx. 1000 rpm for whole cells will separate them from the culture supernatant. Higher speeds will be needed to deposit cell fragments. The cellular pellets are discarded or extracted depending on where the required component is located. Plant cell cultures are often used for production of specific biochemicals such as flavinoids and alkaloids used in foods and chemical production. Culture of plant cells for agrochemical uses is another obvious application.

POTTER HOMOGENIZER
A device for breaking open cells, e.g. plant cells, by using the rotation of a tightly fitting piston inside a glass cylinder.

2.12.3 Mammalian cell culture

There are many different cell lines of human, mammalian, avian and even insect origin. For simplicity, a single type, hybridomas, will be used as an example here. Hybridoma cell lines are prepared using cells from mouse spleen which secrete desired antibodies crossed with a continuous cell line grown *in vitro*. The cells are placed in suspension culture in an incubator shaker at 37°C and 60–100 rpm speed. The cells are transferred to a specially adapted fermentor. Large, broad-bladed impellors are used to minimize shear forces and typical stirrer speeds are 20–100 rpm. A gas mixing station can provide an air/carbon dioxide mixture and CO_2 gas can be used for pH control via a solenoid valve as replacement for a peristaltic pump. Oxygen and nitrogen can be introduced for dissolved oxygen control. For many cell lines, the control of foam is vital as its presence can contribute to cell damage.

HYBRIDOMA CELL LINE
A cell derived from the artificial fusion of a normal, e.g. an antibody-producing cell line from the spleen, with a transformed (immortal) cell line, e.g. a myeloma. The resulting hybrid is immortal and can produce a specific antibody if the correct spleen cell was selected after challenge of an animal with a specific antigen.

SPIN FILTER
This device is normally attached to the drive shaft of an animal cell fermentor and allows culture liquid to be removed while leaving the cells inside the fermentor. It allows for slow growing cells to produce antibodies, for example, over a long period by regular harvesting of culture supernatant and replenishment by a controlled medium feed.

The cells can sometimes grow freely in suspension but more often are anchored to a mesh, alginate beads or a hollow-fibre system. A special mesh can be connected to the drive shaft to provide a 'spin-filter', or to the sample pipe for taking samples of culture fluid relatively cell-free. Some alternative approaches to growth of mammalian cells are as follows:

- Air lift. There is no drive motor and it is the pumping of liquid around the draught tube which provides for mixing and gas exchange.
- Hollow-fibre cartridges. They remain outside the fermentor vessel but culture fluids can circulate between the fibres and the vessel. All mixing and gas exchange is done in the vessel and the enriched medium passes to the cells anchored to the hollow fibres via a pump.
- A gassing basket inside a stirred fermentor. Where cells need good levels of dissolved oxygen but are very sensitive to shear, a large amount of thin-walled silicone tubing supported on a frame can be placed in the vessel to provide a large area for gas exchange. Air is bubbled through the pores in the silicone tubing to give good gas transfer rates without forming large bubbles.

Applications of mammalian cell culture include monoclonal antibody production, culture of viruses for vaccine manufacture and large quantities of specific cells for medical research.

2.12.4 *Algae*

This category covers a diverse range of microorganisms which differ vastly in their growth conditions and metabolic requirements. Some are halophiles (salt-loving), barophiles (survive under high pressure), thermophiles (grow at high temperatures) and psychrophiles (grow at low temperatures). Most need special culture conditons and so it is best to seek out published literature with specific details. Applications for the use of algae include food additives and chemical production by biotransformation.

Summary

- Fermentors are composed of a number of different components which can be grouped by their functions, i.e. temperature control, speed control, continuous culture accessories, etc. A wide range of peripheral devices can enhance the basic facilities of the fermentor, e.g. feed pumps, exit gas analysers, etc.

- Fermentor instrumentation can be of several different types, depending on the age, cost and sophistication of the fermentor. Microprocessor-based systems offer advantages in terms of the flexibility of control and the ability to store operational protocols.

- Many different types of vessels exist such as air lift, fluidized bed, hollow fibre and specially modified stirred tank reactors for containment or *in situ* sterilization.

- Practical steps for autoclaving, use, inoculation and sampling ensure safe and reliable operation for any make or type of fermentor.

- Routine maintenance of electrodes, vessel glass, etc. is repaid in longevity of components and safe operation.

- The range of organisms which can be adapted for growth in fermentors is wide and includes microbes, plant and animal cells.

Suggested reading

Collins, C.H. and Beale, A.J. (eds) (1992) *Safety in Industrial Microbiology and Biotechnology*. London: Butterworth-Heinemann.

Freshney, R.I. (1994) *Culture of Animal Cells: A Manual of Basic Techniques*. New York: A.R. Liss.

Stafford, A. and Warren, G. (eds) (1991) *Plant Cells and Tissue Culture*. Buckingham: Open University Press.

Winkler, M.A. (ed.) (1990) *Chemical Engineering Problems in Biotechnology*. London: SCI/Elsevier.

Wiseman, A. (ed.) (1983) *Principles of Biotechnology*. Surrey: Surrey University Press.

3 Microbiology of Industrial Fermentation

E.M.T. El-Mansi

3.1 Introduction

Microorganisms play a central role in the production of a wide range of industrial chemicals, enzymes and antibiotics. The diversity of microbial processes may be attributed to many factors, including the high surface-to-volume ratio and adaptability to a wide spectrum of carbon and nitrogen sources. The high surface-to-volume ratio supports a very high rate of metabolic turnover, e.g. the yeast *Saccharomyces cereviciae* has been reported to be able to synthesize protein several orders of magnitudes higher than plants. On the other hand, the ability of microorganisms to adapt to different metabolic environments makes them capable of utilizing inexpensive renewable resources such as wastes and by-products of the farming and petrochemical industries as the primary carbon source. Industrially important microorganisms include bacteria, yeasts, moulds and actinomycetes.

The rate of product formation in a given industrial process, a significant parameter, is directly related to the rate of biomass formation which, in turn, is influenced directly or indirectly by a whole host of different environmental factors, e.g. oxygen supply, pH, temperature, accumulation of inhibitory intermediates, etc. It is, therefore, important that we are able to describe growth and production in quantitative terms. The study of growth kinetics and growth dynamics involves the formulation and use of differential equations (for more details see Chapter 4) and while the mathematical derivation of these equations is beyond the scope of this chapter, it is important that we understand how such equations can be used to further our understanding of microbial growth in general and its impact on product formation (yield) and recovery in particular.

While growth kinetics focuses on the measurement of growth rates during the course of fermentation, growth dynamics relates the changes in population (biomass) to changes in growth rate and other parameters, e.g. pH, temperature. To unravel such intricate inter-relationships, the description of growth kinetics and growth dynamics relies on the use of differential equations.

3.2 The growth cycle

Now let us consider the basic equation used to describe microbial growth:

$$\frac{dx}{dt} = \frac{ax}{b} \tag{3.1}$$

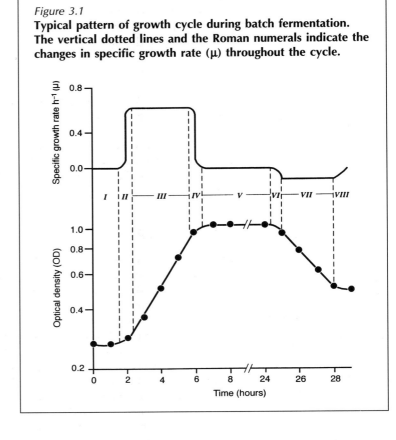

Figure 3.1
Typical pattern of growth cycle during batch fermentation. The vertical dotted lines and the Roman numerals indicate the changes in specific growth rate (μ) throughout the cycle.

Equation (3.1) implies that the rate of biomass (x) formation changes as a function of time (t) and that the rate of change is directly proportional to the concentration of a particular factor (a) such as growth substrate or temperature but is inversely proportional to the concentration of another factor (b) such as inhibitors. In the equation above a and b are independent of time t, and the proportionality factor in equation (3.1) can in effect be ignored. In the early stages of any fermentation process, the increase in biomass is unrestricted and, as such, the pattern of growth follows an autocatalytic first-order reaction (autocatalytic growth) up to a point where either side of equation (3.1) becomes negative, resulting in autocatalytic death.

During batch fermentation, a typical pattern of growth curve, otherwise known as the growth cycle, is observed (Figure 3.1). Clearly a number of different phases of the growth cycle can be differentiated. These are:

I   Lag phase
II Acceleration phase

The term autocatalytic growth is generally used to indicate that the rate of increase in biomass formation in a given fermentation is proportional to the original number of cells present at the beginning of the process, thus reflecting the positive nature of growth

III   Exponential (logarithmic) phase
IV   Deceleration phase
V   Stationary phase
VI   Accelerated death phase
VII   Exponential death phase
VIII   Survival phase

The changes in the specific growth rate (μ) as the organism progresses through the growth cycle can also be seen in Figure 3.1.

We shall now describe the metabolic events and their implications in as far as growth, survival and productivity are concerned. Naturally the scenario begins with the first phase of the growth cycle.

3.2.1   The lag phase

In this phase the organism is simply faced with the challenge of adapting to the new environment. While adaptation to glucose as a sole source of carbon appears to be relatively simple, competition with other carbon sources, although complex, is resolved in favour of glucose through the operation of two different mechanisms, namely catabolite inhibition and catabolite repression.

Adaptation to other carbon sources, however, may require the induction of a particular set of enzymes which are specifically required to catalyse transport and hydrolysis of the substrate, e.g. adaptation to lactose, or to fulfil anaplerotic as well as regulatory functions as is the case in adaptation to acetate as the sole source of carbon and energy. Irrespective of the mechanisms employed for adaptation, the net outcome at the end of the lag phase is a cell that is biochemically vibrant, i.e. capable of transforming chemicals to biomass, i.e. growth.

Entering a lag phase during the course of industrial fermentation is not desirable as it is very costly and as such should be avoided. The question of whether a particular organism has entered a lag phase in a given fermentation process can be determined graphically by simply plotting log n (biomass) as a function of time, as shown in Figure 3.2.

Note that the transition from the lag phase to the exponential phase is generally interrupted by an acceleration phase, as described earlier (Figure 3.1). This difficulty can be easily overcome by extrapolating the lag phase sideways and the exponential phase downwards as shown, with the point of interception (L) taken as the time at which the lag phase ended. What is also interesting about the graph in Figure 3.2 is that if one continues to extrapolate the exponential phase downwards, then the point at which the ordinate is intersected gives the number of cells that were viable and metabolically active at the point of inoculation.

The term growth cycle is used to describe the overall pattern displayed by microorganisms during growth in batch cultures. It is noteworthy that such a cycle is by no means a fundamental property of the bacterial cell, but rather a consequence of the progressive decrease in food supply or accumulation of inhibitory intermediates in a closed system to which no further additions or removals are made.

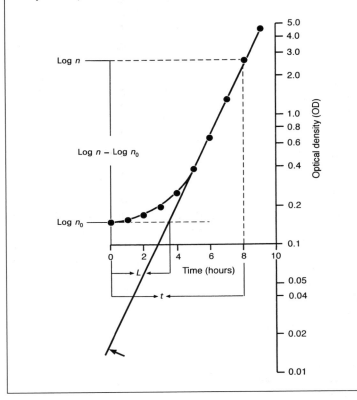

Figure 3.2
Graphical determination of the lag phase and the number of viable cells at the onset of batch fermentation. As the exponential phase is extrapolated downwards, it intercepts the extrapolated line of the lag phase and the ordinate respectively (see text for details).

The question of whether a lag (L) has occurred during the course of the fermentation process and for how long can be easily determined from the following equation:

$$\frac{\log n - \log n_0}{t - L} = \frac{\log 2}{T} \qquad (3.2)$$

where n is the total number of cells after a given time (t) since the start of fermentation, n_0 is the number of cells at the beginning of fermentation and T is the organism's mean generation time (doubling time). Equation (3.2) describes the exponential growth, taking into consideration a lag phase in the process. In the following section we shall describe the exponential phase in general and the derivation of equation (3.2) in particular.

CATABOLITE
INHIBITION AND
CATABOLITE
REPRESSION
In the presence of two
sugar substrates,
microorganisms such as
E. coli employ catabolite
inhibition and repression
to ensure preferential
utilization and
metabolism of glucose
over its counterpart.
While the former
mechanism is exerted at
the level of uptake and
ensures inhibition of
transport of the
competing sugar, e.g.
lactose or citrate, the
latter mechanism is
exerted at the level of
transcription, i.e.
prevents the formation of
m-RNA, and involves
cyclic adenosine
monophosphate (cAMP)
and catabolite repressor
protein (CRP). Recent
investigations have also
revealed the presence of
catabolite repression
mechanisms that are
independent of cAMP
(Saier, 1996).

3.2.2 *The exponential phase*

In this phase, each cell increases in size, and providing that
conditions are favourable, divides into two which, in turn, grow
and divide and the cycle continues. During this phase, the cells
are capable of transforming the primary carbon source into
biosynthetic precursors, reducing power (NAD(P)H, FADH$_2$,
HS-CoA, etc.) and energy which is generally trapped in the
form of ATP, PEP and proton gradients. Such a complex function is achieved through the activities of various enzymes of central metabolism. The biosynthetic precursors thus generated are
then channelled through various biosynthetic pathways for the
biosynthesis of various monomers (amino acids, nucleotides, fatty
acids, sugars) which, in turn, are polymerized to give the required polymers (proteins, nucleic acids, ribonucleic acids, lipids).
Finally, these polymers are assembled in a precise way and the
cell divides to give the new biomass characteristic of each organism. The time span of each cycle (cell division) is known as
generation time or **doubling time** but because we generally deal
with many millions of cells in bacterial cultures, the term **mean
generation time** is more widely used to reflect the average generation times of all cells in the culture. Such a rate, providing
conditions are favourable, is fairly constant.

If a given number of cells (n_0) are inoculated into a suitable
medium and the organism was allowed to grow exponentially,
then the number of cells after one generation is $2n_0$; at the end
of two generations, the number of cells becomes $4n_0$ (or $2^2 n_0$). It
follows therefore that at the end of a certain number of generations (Z), the total number of cells equals $2^Z n_0$. If the total number
of cells, or its log value, at the end of Z generations is known,
then

$$n = n_0 2^Z \tag{3.3}$$

$$\log n = \log n_0 + Z \log 2 \tag{3.4}$$

To determine the number of cell divisions, i.e. number generations (Z) which have taken place during fermentation, the above
equation can be modified to give:

$$Z = \frac{\log n - \log n_0}{\log 2} \tag{3.5}$$

If T is the mean generation time required for the cells to double
in number and t is the time span over which the population has
increased exponentially from n_0 to n, then

$$Z = \frac{t}{T} = \frac{\log n - \log n_0}{\log 2} \tag{3.6}$$

If during the course of a particular fermentation, a lag time has been demonstrated, equation (3.6) can be modified to take account of this observation. The modified equation is as follows:

$$Z = \frac{t - L}{T} = \frac{\log n - \log n_0}{\log 2} \tag{3.7}$$

The above equations (3.6) and (3.7) can be rearranged to give the familiar equations governing the determination of the mean generation time (T) as follows:

$$\frac{\log n - \log n_0}{t} = \frac{\log 2}{T} \tag{3.8}$$

$$\frac{\log n - \log n_0}{t - L} = \frac{\log 2}{T} \tag{3.9}$$

While equation (3.8) describes the exponential phase of growth in pure terms, equation (3.9) takes into consideration the existence of a lag phase in the process.

The exponential pattern (i.e. 2–4–8–16, and so on) demonstrated in Figure 3.3 obeys equations (3.8) and (3.9) for logarithmic growth. Note the different position of $\log n_0$ to that cited in Figure 3.2.

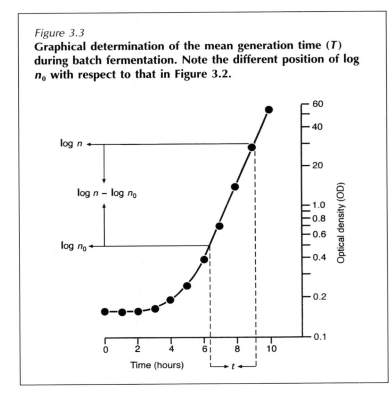

Figure 3.3
Graphical determination of the mean generation time (T) during batch fermentation. Note the different position of log n_0 with respect to that in Figure 3.2.

Although the use of log to the base 2 has the added advantage of being able to determine the number of generations relatively easily, because an increase of one unit in log $2n$ corresponds to one generation, the majority of researchers continue to use log to the base 10. In this case, the slope of the line (Figure 3.3) equals $\mu/2.303$. The relationship between the mean generation time (T) and the specific growth rate (μ) can be described mathematically by the following equation:

$$\ln 2 = \mu T \qquad (3.10)$$

or

$$\mu = \ln 2/T \qquad (3.11)$$

Because $\ln 2 = 0.693$, either of the above equations can be rearranged to give

$$T = \frac{0.693}{\mu} \qquad (3.12)$$

During the course of exponential growth, the culture reaches a steady state and as such the intracellular concentrations of all enzymes, cofactors and substrates are considered to be constant. During this phase, one can therefore safely assume that all bacterial cells are identical and that the doubling time is constant with no loss in cell numbers due to cell death. The rate of growth of a given population (N) represents, therefore, the rate of growth of each individual cell in the population multiplied by the total number of cells. Such a rate can be described mathematically by the following differential equation:

$$\frac{dN}{dt} = \mu N \qquad (3.13)$$

This equation implies that the rate of new biomass formation is directly proportional to the specific growth rate (μ) of the organism under investigation and the number of cells (N). This pattern of growth may be referred to as **autocatalytic**: a term which has been described earlier. If the above equation describes the exponential phase correctly, then a straight line should be obtained when $\ln N$ is plotted as a function of time (t). During the exponential phase, it is generally assumed that all cells are identical and as such the specific growth rate (μ) of individual cells equals that of the whole population. Equation (3.13) can therefore be rearranged to take account of the fact that dN/dt is proportional to the number of cells (N) and that the specific growth rate (μ) is a proportionality factor to give:

$$\mu = \frac{1}{N}\frac{dN}{dt} \qquad (3.14)$$

Doubling time or mean generation time (T) is the time required for a given population (N_0) to double in number ($2N_0$).

The specific growth rate (μ) is usually expressed in terms of units per hour (h^{-1}).

Although equations (3.9) and (3.10) describe the exponential phase satisfactorily, in some fermentations, as is the case during growth of *E. coli* on sodium acetate (M. El-Mansi, unpublished results), the organism fails to maintain a steady state for any length of time and so μ falls progressively with time until the organism reaches the stationary phase. Such a drop in growth rate (μ) can be accounted for by a whole host of different factors including nutrient limitations, accumulation of inhibitors and/or a crowding factor, i.e. as the population increases in size μ decreases.

3.2.2.1 Metabolic interrelationship between nutrient limitations and specific growth rate (μ)

To account for the effect of nutrient limitations on growth rate (μ), Monod modified the above equation so that the effect of substrate concentration (limitations) on μ could be assessed quantitatively. Monod's equation is as follows:

$$\mu = \frac{\mu_m S}{K_S + S} \qquad (3.15)$$

where S is the concentration of the limiting substrate, μ_m is the maximum specific growth rate and K_S is the saturation constant, i.e. when $S = K_S$ then $\mu = \mu_m/2$ as illustrated in Figure 3.4.

During growth in a steady state, the following equations may be used to describe growth using biomass (X) rather than number of cells (N) as a measure of growth:

$$\frac{dX}{dt} = \mu X \qquad (3.16)$$

$$\frac{dS}{dt} = -\mu \frac{X}{Y} \qquad (3.17)$$

where X is the biomass concentration, Y is the growth yield (g biomass generated per gram substrate utilized). It is noteworthy, however, that during steady state growth, i.e. in a chemostat or a turbidostat, the terms used to describe the bacterial numbers (N) and biomass (X) in equations (3.16) and (3.17) are identical. This is not necessarily the case in batch cultures because the size and shape of bacterial cells vary from one stage of growth to another.

While equation (3.15) addresses the relative change in biomass (the first variable) with respect to time, equation (3.16) addresses the relative change in substrate concentration (the second variable) as a function of time. Note that following the exhaustion

Figure 3.4

Graphical determination of the maximum specific growth rate (μ) and the saturation constant (K_s). Note that this graph was constructed without taking maintenance energy into consideration. With maintenance in mind the curve should slide sideways to the right in direct proportion to the fraction of carbon diverted towards maintenance.

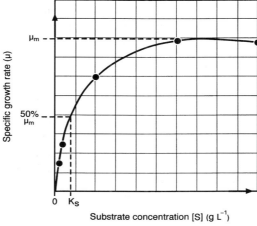

of substrate $dX/dt = dS/dt = 0$. While either equation (3.16) and (3.17) can surely predict the deceleration and stationary phases of growth, it should be remembered that nutrient limitation is not the only reason for the deceleration of growth and subsequent entry into the stationary phase. In addition to environmental factors, crowding is reported to have an adverse effect on growth rate (μ), i.e. growth rate diminishes as the size of the population increases. Although the equation describing the effect of crowding on growth rate (Verhulst–Pearl logistic equation) predicts a sigmoidal pattern of growth and fits rather well with the growth curve for populations of higher organisms, yeasts as well as bacteria, it will not be further discussed in this chapter as it overlooks the effect of other environmental factors on growth.

3.2.3 *Stationary phase and cell death*

As the exponential phase draws to an end, the organism enters the stationary phase and then the death phase. The equations used earlier to describe growth do not adequately describe cell death. However, if we assume that the kinetics of death are similar to that of growth, then the specific rate of cell death (λ) can be described mathematically by the following equation:

$$\frac{dN}{dt} = (\mu - \lambda)N \hspace{4cm} (3.18)$$

Equation (3.18) clearly indicates that if μ is greater than λ, then the organism will grow at a rate equal to $\mu - \lambda$. If on the other hand μ is less than λ, then the population dies at a rate equal to $\lambda - \mu$. Under conditions where μ equals λ, neither growth nor death is observed: a situation which is thought to prevail throughout the course of the stationary phase. As the energy supply continues to fall, the equilibrium between λ and μ will finally shift in favour of λ and consequently cell death begins.

Microorganisms respond differently to nutrient limitations during the stationary phase. For example, while *Bacillus subtilis* and other Gram-positive, spore-forming organisms respond by sporulation, *E. coli* and other Gram-negative, non-spore-forming bacteria cannot respond in the same way and as such other mechanisms must have evolved. Although a fraction of the bacterial population die during this phase of growth, a relatively large number remains viable for a long time despite starvation. The ability of cells to remain viable despite a prolonged period of starvation is advantageous, as most microorganisms in Nature are subject to nutrient limitations in one form or another. Our understanding of the molecular mechanisms employed by microorganisms for survival during this phase of growth is rather limited. However, contrary to the notion that microorganisms enter a logarithmic death phase soon after the onset of the stationary phase, recent investigations have revealed that some microorganisms such as *Saccharomyces cerevisiae* and *E. coli* adapt to starvation, presumably through mutations and induction mechanisms, very effectively.

The ability of some cells to survive prolonged starvation inspired some researchers to pursue the question of whether such cells are biochemically distinguishable from their predecessors which were actively growing or had just entered the stationary phase. In the case of *Saccharomyces cerevisiae*, analysis of mRNAs which are specifically expressed following the onset of the stationary phase revealed the presence of a new family of genes. This family is referred to as SNZ (short for snooze) and sequence analysis revealed that they are highly conserved. The functional role of each member of this family, however, remains to be determined. It is interesting that some researchers use the term **Viable But Non-Culturable** (VBNC) to describe cells in the stationary phase. However, as colony-forming ability is our only means of assessing whether a particular cell is alive or not, other microbiologists argue the case for **reversibility**, i.e. the cell's ability to transform itself from being dormant to being metabolically active. Recent investigations have revealed that resuscitation of stationary phase cells may be aided by the excretion of pheromones; this has recently been demonstrated in *Micrococcus luteus*.

In the case of *E. coli*, an attempt was made to identify the genes which are uniquely turned on in the stationary phase in response to starvation, i.e. not expressed during the exponential phase, with the aid of random transposon mutagenesis. A number of mutants which were unable to survive in the stationary phase were isolated and, subsequently designated as **survival negative mutants** (Sur⁻) and the genes involved were designated as *sur* genes. While some genes, e.g. *surA*, are required for survival during 'famine' (starvation), others, e.g. *surB*, enable the organism to exit the stationary phase as conditions change from 'famine' to 'feast'. Recent studies have also revealed that starvation induces the production of a stationary phase-specific transcriptional activator (sigma factor) that is essential for the transcription of *sur* genes by RNA polymerase. The physiological function of the *sur* gene products is to downshift the metabolic demands made on central and intermediary metabolism for maintenance energy.

3.2.4 *Maintenance and survival*

Maintenance can be defined as the minimal rate of energy supply that is required to maintain the viability of a particular organism without contributing to biosynthesis. The fraction of carbon oxidized in this way is, therefore, expected to end up in the form of carbon dioxide (CO_2). The need for maintenance energy is obvious as the 'living-state' of any organism, including ourselves, is remote from equilibrium and as such demands energy expenditure. Moreover, in addition to the carbon processed for maintenance, another fraction of the carbon source may be wasted through excretion, e.g. acetate, α-ketoglutarate, succinate, lactate, etc. In this context, excretion of metabolites ought to be seen as an accidental consequence of central metabolism rather than by design to fulfil certain metabolic functions. The fraction of carbon required for maintenance differs from one organism to another and from one substrate to another. For example, maintenance requirements for the lactose phenotype of *E. coli* are greater than those observed for the glucose phenotype as the *lac*-permease is much more difficult to maintain than the uptake system employed for the transport of glucose, i.e. the phosphotransferase system (PTS).

The metabolic interrelationship between maintenance and growth was first described by Pirt (1965). In his theoretical treatment of this aspect, he formulated a maintenance coefficient (m) based on the earlier work of Monod, and defined the coefficient as the amount of substrate consumed per unit mass of organism per unit time (e.g. g substrate per g biomass per h). If s represents the energy source (substrate), then the rate of substrate consumption for maintenance is governed by the following equation:

$$-(ds/dt)_M = mX \qquad (3.19)$$

where m is the maintenance coefficient. The yield can therefore be related to maintenance by the following equation:

$$Y = \frac{\Delta x}{(\Delta s)_G + (\Delta s)_M} \qquad (3.20)$$

where Y is the growth yield, Δx is the amount of biomass generated, $(\Delta s)_G$ is the amount of substrate consumed in biosynthesis and $(\Delta s)_M$ is the amount of substrate consumed for maintenance of cell viability.

If maintenance energy $(\Delta s)_M$ was determined and found to be zero, then equation (3.20) reduces to:

$$Y_G = \frac{\Delta x}{(\Delta s)_G} \qquad (3.21)$$

Equation (3.21) clearly means that growth yield (Y) is a direct function of the amount of substrate utilized. The growth yield (Y, g dry weight per g substrate) obtained in this case may be referred to as the **true growth yield** to distinguish it from the Y_G where maintenance $(\Delta s)_M$ is involved. However, if a specific fraction of carbon is diverted towards maintenance energy or survival, then one might conclude that the slower the growth rate (μ), the higher the percentage of carbon that is diverted towards maintenance, the less the growth yield.

On the other hand, however, during unrestricted growth, i.e. no substrate limitation, the interrelationship between substrate concentration and maintenance can be assessed from the following equation (Pirt, 1965):

$$ds/dt = (ds/dt)_M + (ds/dt)_G \qquad (3.22)$$

In this case, equation (3.22) implies that the overall rate of substrate utilization equals the rates of substrate utilization for both maintenance and growth. With growth rate expressed in the usual way, i.e. $dx/dt = \mu x$, and assuming that Y equals Y_G, equations (3.16) and (3.18) can be rearranged to give:

$$1/Y = m/\mu + 1/Y_G \qquad (3.23)$$

According to equation (3.23), if m and Y_G are constants, then the plot of $1/Y$ against $1/\mu$ should give a straight line, the slope of which is maintenance (m) and the point at which the ordinate is intercepted is equal to $1/Y_G$, as illustrated in Figure 3.5.

During the course of fermentation, the minimum rate of energy supply may fall below maintenance requirements due to a shortfall in the supply of phosphorylated intermediates, ATP

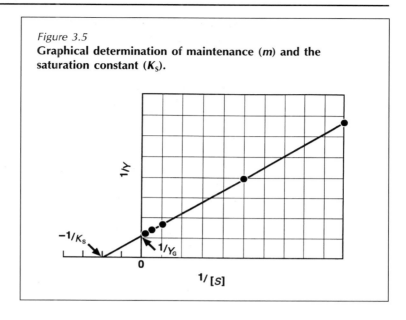

Figure 3.5

Graphical determination of maintenance (*m*) and the saturation constant (*K*ₛ).

and/or reducing powers (NAD(P)H, FADH$_2$, etc.). Although such a drop in energy supply may not be fatal, it is likely to have an adverse effect on the fitness of the organism, i.e. the cells become less capable of taking advantage of favourable changes in the environment and consequently cells resume growth after a lag period. On the other hand, if the cells were able to grow immediately following inoculation or transfer into another medium, i.e. without lag, then the cells must have been left in optimal conditions. However, if the cells were to enter the lag phase prior to growth then it is fair to suggest that the organism was left in a sub-optimal state, i.e. the organism must have suffered a drop in energy supply below its maintenance requirements.

It follows, therefore, that the physiological state of the cell at the time when energy supply falls below maintenance is very important and that the deeper the shortfall in energy supply, the longer the lag period. A lag period due to the above condition is surely different to that observed when the cells are faced with the challenge of changing phenotype, as is the case when glucose phenotype of *Escherichia coli* is forced to change to lactose phenotype. Individual cells in a given population may respond differently when a given drop of energy supply is exerted. The ability of a given population to respond successfully to 'shift-up' or to survive a 'shift-down' in nutrients appears to be directly related to the intracellular concentration of ribosomes. For example, in comparison with cells grown in rich medium, restricted growth of *Salmonella typhimurium* was accompanied by a 35-fold drop in the concentration of ribosomes. In the yeast

Ribosome particles consist of protein and RNA. They are the target to which mRNA and amino acyl-tRNA must bind in preparation for translation (the conversion of mRNA to a polypeptide chain).

Autogenous regulation is a regulatory control mechanism which is exerted at the level of translation, i.e. the conversion of mRNA to protein, rather than transcription.

Saccharomyces cerevisiae, a drop in growth rate from $0.40\ h^{-1}$ to $0.10\ h^{-1}$ was accompanied by a drop in the cellular concentrations of RNA, DNA and proteins from 12.1%, 0.6% and 60.1% to 6.3%, 0.4% and 45% respectively (Nissen *et al.*, 1997). Furthermore, it was demonstrated that the cells were capable of 'scaling up' or 'scaling down' the intracellular level of ribosomes in response to the 'shift-up' or the 'shift-down' in nutrients respectively.

Recent investigations have revealed that such ability to modulate the cellular content of ribosomes is achieved through a mechanism of autogenous regulation.

To account for such a phenomenon, Nomura *et al.* (1984) proposed a new hypothesis, **translational couplings**, which simply implies that binding to and initiation of translation at the first ribosome binding site of the operon is essential in exposing the ribosome binding sites of all other cistrons and that inhibition or prevention of translation at the first binding site means that all other sites within the polycistronic message will remain unexposed and as such untranslated.

The argument over the question of whether smaller cells are less able to cope with environmental changes, i.e. nutrient limitations and the drop in energy supply below maintenance, because they contain less ribosomes can now be answered to the satisfaction of everyone as recent research revealed that it is not the number of ribosomes that matter but rather their concentration inside the cell. Furthermore, the need for the cell size to be relatively large during growth under no limitations is not a reflection of high concentration of ribosomes but rather the need to attain a certain mass before replication of the chromosome can be initiated (Donachie, 1968). The essence of Donachie's discovery is that if initiation of replication is triggered in response to a certain volume of cell mass, then the faster the cell divides, the faster the growth rate, the larger the size of the cell.

3.3 Diauxic growth

The ability of microorganisms to display biphasic (diauxic) growth patterns (Figure 3.6) is well documented. For example, *E. coli* and *Aerobacter aerogenes* display such a pattern when faced with two different substrates, glucose and lactose in the case of the former (Figure 3.6) and glucose and citrate in the case of the latter. This is generally due to the preferential utilization of glucose to either of the competing substrates. This can be explained on the grounds that glucose has been found to exert catabolite inhibition, i.e. inhibition of other sugars or organic acids uptake systems, and catabolite repression of the enzymes

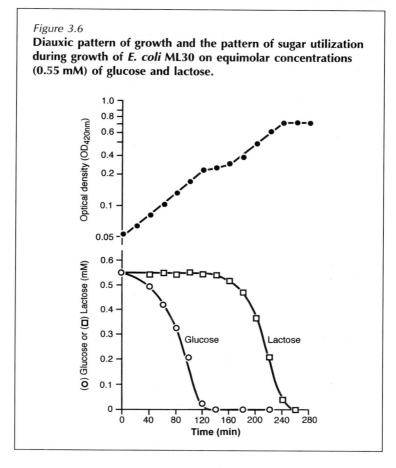

Figure 3.6

Diauxic pattern of growth and the pattern of sugar utilization during growth of *E. coli* ML30 on equimolar concentrations (0.55 mM) of glucose and lactose.

required for the transport as well as hydrolysis of other sugars, as is the case in lactose metabolism.

3.4 Growth yield

The relative contribution of each substrate to total biomass formation can be assessed with a good deal of precision providing that growth limitation is only enforced as a result of exhaustion of the substrate in question rather than the accumulation of toxic end-products or adverse changes in pH. The relationship between total biomass formation and the concentration of any given substance can therefore be determined from the plot of log n (biomass) as a function of substrate concentration and providing that all other components are in excess, the organism will enter the stationary phase upon depletion of the substrate or the substance in question. This experiment should obviously be done over a wide range of substrate concentrations and from the slope obtained, the **yield constant** or **yield coefficient** per unit of the substance in question can be determined. If the unit is, say, one mole of a substrate, then the yield coefficient obtained is referred to as the **molar growth yield coefficient**. This method is used widely as a biological assay for the determination of vitamins, amino acids, purines and pyridines.

3.5   Fermentation balances

3.5.1 Carbon balance

Apart from the fraction of carbon used for biomass formation, the remainder is partitioned between products and by-products, including carbon dioxide. The ratio of recovered carbon to that present at the onset of fermentation is referred to as the **carbon balance** or **carbon recovery index**. Such a balance or an index is a measure of efficiency; i.e. the higher the index, the higher the carbon recovery, the more efficient the fermentation process. The carbon balance is generally calculated by working out the number of moles (or millimoles) produced of a given product per 100 moles (or millimoles) of substrate utilized. The values obtained can then be multiplied by the number of carbon atoms in each respective molecule. The resulting values for the products in question can then be totalled and compared with that of the substrate. If the values are equal, i.e. 1 : 1 ratio, then a complete recovery of carbon into product formation has been achieved. Although this is theoretically possible, our experience indicates otherwise as part of this carbon is used for maintenance and assimilation to support growth, however slow. A complete carbon balance can, therefore, be calculated for any given fermentation if the fraction of carbon diverted towards biosynthesis and maintenance is determined. In addition to the carbon balance, some fermentations involve calculation of the redox balance.

3.5.2 Redox balance

Fermentations of sugars and other primary carbon sources give rise to a whole host of different intermediates, some of which are phosphorylated (energy rich) while others are not. While biosynthetic intermediates are utilized for the biosynthesis of monomers, other intermediates are produced in excess and this is balanced by their excretion into the medium. The stoichiometry of product and by-product formations of any given fermentation process can be ascertained by carefully analysing the culture filtrates at different stages (for an extensive treatment of this aspect see Chapter 6). From a physiological standpoint and in order for fermentation to go to completion, the redox balance must be maintained.

3.6   Efficiency of central metabolism

The efficiency of carbon conversion to biomass and desirable end-products is influenced by many different factors (see Chapters 4, 6 and 7 for more details) of which futile cycling is a major contributor. For example, *Klebsiella aerogenes* has been shown to have an NAD$^+$-dependent pyruvate reductase, an enzyme which catalyses the formation of D-lactate, and an FAD-dependent D-lactate dehydrogenase which catalyses the conversion of D-lactate to pyruvate. The operation of these two enzymes provides a futile cycle in which NADH,H$^+$ is oxidized at the expense of reducing FAD, thus bypassing the first phosphorylation site in the electron transport chain (Figure 3.7). It follows therefore that any changes in the metabolic environment which lead to a transient increase in the intracellular level of pyruvate may lead to the operation of this futile cycle and, in turn, lower the efficiency of central metabolism. The presence and function of glutamine synthetase and glutaminase in ammonia-limited cultures represents yet another possible futile cycle (Figure 3.8) in which the synthetase generates glutamate while the other affects its hydrolysis with a net loss of one molecule of ATP per turn (Tempest, 1978).

In addition to futile cycling, flux to metabolite excretion in general and acetate in particular is another significant factor in influencing the efficiency of central metabolism (El-Mansi and

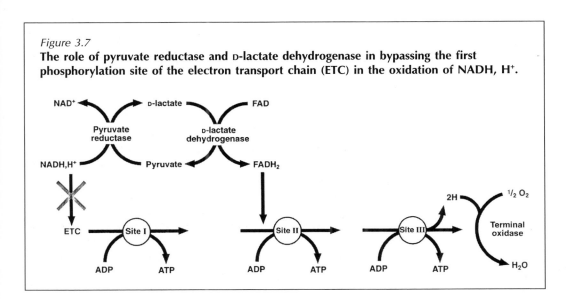

Figure 3.7

The role of pyruvate reductase and D-lactate dehydrogenase in bypassing the first phosphorylation site of the electron transport chain (ETC) in the oxidation of NADH, H$^+$.

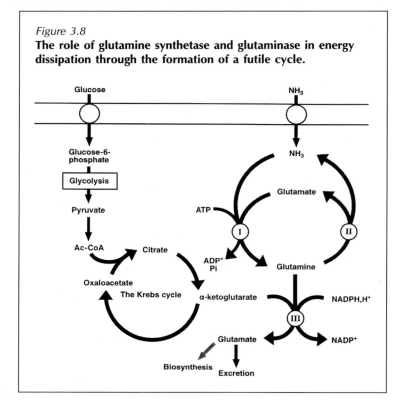

Figure 3.8
The role of glutamine synthetase and glutaminase in energy dissipation through the formation of a futile cycle.

Holms, 1989). Furthermore, because acetate, particularly in its undissociated form, has proved to be a potent uncoupler of oxidative phosphorylation, any conditions which promote flux to acetate excretion will, in turn, diminish flux to product formation due to the diversion of a much larger fraction of carbon towards high maintenance energy requirements (El-Mansi, unpublished results).

Summary

- The diverse array of metabolic networks, together with a very high rate of metabolic turnover, makes microorganisms ideal candidates for the production of fine chemicals using fermentation technology.

- Microorganisms change in size as the growth rate changes: the faster the growth rate, the larger the size of the cell, which is consistent with the view that initiation of replication is dependent on cell mass.

- Microorganisms change in composition as a function of growth rate, e.g. while the relative concentrations of protein, DNA and

RNA in *Saccharomyces cerevisiae* increase (in ascending order) with growth rate, the intracellular concentration of glycogen, trehalose and carbohydrates decrease.

- The drop in energy supply below maintenance is not necessarily fatal but may lead to lag, the severity of which is directly proportional to its magnitude and the relative concentration of ribosomes within the cell.

- The question of whether a particular organism has lagged following inoculation, together with the number of viable cells at the onset of fermentation, can be ascertained graphically.

- The interrelationships among various growth parameters during unrestricted and restricted growth have been described.

- Microorganisms adopt different tactics to survive starvation. While Gram-positive organisms resort to sporulation, *Saccharomyces cereviciae* and *E. coli* induce specific genes, the products of which are essential for survival and for exiting the stationary phase.

Acknowledgement

The author wishes to thank Dr Bill Robin, Department of Mathematics, Napier University, Edinburgh, for his critical and thorough revision of the mathematical equations reported in this chapter.

References

Donachie, W.D. (1968) Relationship between cell size and time of initiation of DNA replication, *Nature*, **219**, 1077–1083.

El-Mansi, E.M.T. and Holms, W.H. (1989) Control of carbon flux to acetate excretion during growth of *Escherichia coli* in batch and continuous cultures, *J. Gen. Microbiol.*, **135**, 2875–2883.

Nissen, L.N., Schulze, U., Nielsen, J. and Villadsen, J. (1997) Flux distribution in anaerobic, glucose-limited continuous cultures of *Saccharomyces cerevisiae*, *Microbiology*, **143**, 203–218.

Nomura, M., Gourse, R. and Baughman, G. (1984) Regulation of synthesis of ribosomes and ribosomal components, *Ann. Rev. Biochem.*, **53**, 75–89.

Pirt, S.J. (1965) The maintenance energy of bacteria in growing culture, *Proc. R. Soc. London, Ser. B*, **163**, 224–231.

Saier, M.H. Jr (1996) Cyclic AMP-independent catabolite repression in bacteria, *FEMS Microbiol. Lett.*, **138**, 97–103.

Tempest, D.W. (1978) The biochemical significance of microbial growth yield, *Trends Biochem. Sci.*, **3**, 180–184.

Suggested reading

Bazin, M.J., Wood, A.P. and Paget-Brown, D. (1997) Analysis of microbial growth data, in Rhodes, P.M. and Stanbury, P.F. (eds) *Applied Microbial Physiology*, Oxford, New York, Tokyo: IRL Press, pp. 103–211.

Dawes, E.D. and Ribbons, D.W. (1964) Endogenous metabolism of bacteria, *Bacteriol. Rev.*, **28**, 126–141.

Neidhardt, F.C., Ingraham, J.L. and Schaechter, M. (1990) *Physiology of the Bacterial Cell: A Molecular Approach*, Sunderland, MA: Sinauer Associates.

4 Fermentation Kinetics

Jens Nielsen

4.1 Introduction

Growth of cell populations is the result of many chemical reactions leading to synthesis of macromolecules such as DNA, RNA and proteins. Prior to cell division, the cells increase in size (or extend their hyphae in the case of filamentous microorganisms) as the macromolecules are assembled en route to biomass formation. The growth of cell populations can be quantified by many different methods. One can measure the total number of cells – both living and dead – using a cell counter or by direct counting in a diluted cell suspension using microscopy. Alternatively one can determine the total dry weight (see Box 4.1) or the cellular content of RNA, DNA or protein. Other methods are based on the measurement of turbidity, which is generally linearly related to the dry weight determination, and today there are a number of commercially available sensors that allow *in situ* measurements of the turbidity (Olsson and Nielsen, 1997).

Growth of cell populations (see Chapter 3 for more details) is often illustrated with a batch-wise growth of a unicellular organism (either a bacterium or a yeast). Here the growth occurs

Box 4.1 **Standard operating procedure (SOP) for dry weight determination**

Biomass is most frequently determined by dry weight measurements. This can be done either using an oven or a microwave oven, with the latter being the fastest procedure. An important prerequisite for the measurement is that the sample is dried completely, and it is therefore important to apply a consistent procedure. A suggested protocol is as follows:

1. Dry the filter (pore size 0.45 μm for yeast or fungi, 0.20 μm for bacteria) on a glass dish in the microwave oven on 150 W for 10 min. Place a tissue paper between the glass and the filter so that the filter does not stick to the glass.

2. Place the filter in a desiccator and allow to cool for 10–15 min. Weigh the filter.

3. Filter the cell suspension through the filter and wash the cells with demineralized water.

4. Place the filter on the glass dish again and dry in the microwave oven for 15 min at 150 W.

5. Put the filter in a desiccator and allow to cool for 10–15 min. Weigh the filter.

6. If more than 30 mg dry weight is present on the filter, the time in the microwave oven may have to be longer.

in a constant volume of medium with one growth-limiting substrate component that is used by the cells. Cell growth is generally quantified by the so-called **specific growth rate** $\mu(h^{-1})$, which for such a culture is given by:

$$\mu = \frac{1}{x} \frac{dx}{dt} \tag{4.1}$$

where x is the biomass concentration (or cell number). The specific growth rate is related to the **doubling time** $t_d(h)$ of the biomass through:

$$t_d = \frac{\ln 2}{\mu} \tag{4.2}$$

The doubling time t_d is equal to the generation time for a cell, i.e. the length of a cell cycle for unicellular organisms, which is frequently used by life scientists to quantify the rate of cell growth.

In the design and optimization of a fermentation process a quantitative description of the process is required. Considering the complex nature of microbial growth, a quantitative description of the fermentation process can be a very difficult task. Furthermore, often the product is not the cells themselves but a compound synthesized by the cells, and depending on the type of product the kinetics of its formation may vary from one phase of growth to another. Thus, while primary metabolites are typically formed in conjunction with cellular growth, an inverse relationship between product formation and cell growth is often found in the case of secondary metabolites, and here the productivity may be greatest in the stationary phase.

With these differences it is clear that quantification of product formation kinetics may be a difficult task. However, with the rapid progress in biological sciences our understanding of cellular function has increased dramatically in recent years, and this may form the basis for far more advanced modelling of cellular growth kinetics than seen earlier. Thus, in the literature one may find mathematical models describing events like gene expression, kinetics of individual reactions in central pathways, together with macroscopic models that describe cellular growth and product formation with relatively simple mathematical expressions. These models cannot be compared directly because they serve completely different purposes, and it is therefore important to consider the aim of the modelling exercise in a discussion of mathematical models.

In this chapter the applications of kinetic modelling to fermentation processes (or cellular processes) will be discussed, and we start this discussion by presenting a framework for kinetic models.

4.2 Framework for kinetic models

The net result of the many biochemical reactions within a single cell is the conversion of substrates to biomass and metabolic end-products (see Figure 4.1). Clearly the number of reactions involved in the conversion of, say, glucose into biomass and desirable end-products is very large, and it is therefore convenient to adopt the structure proposed by Neidhardt *et al.* (1990) for cellular metabolism, which can be summarized as follows:

- **Assembly reactions** carry out chemical modifications of macromolecules, their transport to prespecified locations in the cell, and finally, their association to form cellular structures such as cell walls, membranes, the nucleus, etc.
- **Polymerization reactions** represent directed, sequential linkage of activated molecules into long (branched or unbranched) polymeric chains. These reactions form macromolecules from a moderately large number of building blocks.
- **Biosynthetic reactions** produce the building blocks used in the polymerization reactions. They also produce coenzymes and related metabolic factors, including signal molecules. There is a large number of biosynthetic reactions, which occur in functional units called biosynthetic pathways, each consisting of one to a dozen sequential reactions leading to the synthesis of one or more building blocks. Pathways are easily recognized and are often controlled *en bloc*. In some cases their reactions are catalysed by enzymes made from a single piece of mRNA transcribed from a set of contiguous genes forming an operon. All biosynthetic pathways begin with one of 12 precursor metabolites. Some pathways begin directly with such a precursor metabolite, others indirectly

---

Figure 4.1

The overall result of intracellular biochemical reactions. In addition to the formation of biomass constituents, e.g. cellular protein, lipids, RNA, DNA and carbohydrates, substrates are converted into primary metabolites, e.g. ethanol, acetate, lactate; secondary metabolites, e.g. penicillin; and/or extracellular macromolecules, e.g. enzymes, heterologous proteins, polysaccharides.

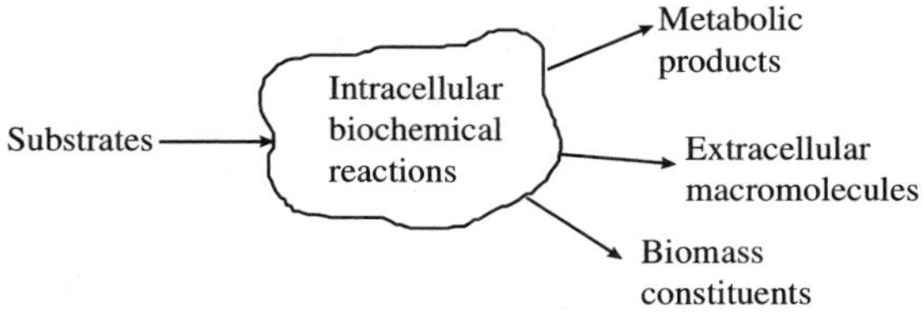

Table 4.1 Overall composition of an average cell of *Escherichia coli*

Macromolecule	Percentage of total dry weight	Different kinds of molecules
Protein	55.0	1050
RNA	20.5	
rRNA	16.7	3
tRNA	3.0	60
mRNA	0.8	400
DNA	3.1	1
Lipid	9.1	4
Lipopolysaccharide	3.4	1
Peptidoglycan	2.5	1
Glycogen	2.5	1
Metabolite pool	3.9	

Data are taken from Ingraham *et al.* (1983).

by branching from an intermediate or an end-product of a related pathway.

- **Fuelling reactions** produce the 12 precursor metabolites needed for biosynthesis. Additionally, they generate Gibbs free energy in the form of ATP which is used for biosynthesis, polymerization and assembling reactions. Finally, the fuelling reactions produce the reducing power needed for biosynthesis. The fuelling reactions include all biochemical pathways referred to as catabolic pathways (degrading and oxidizing substrates).

Thus, the conversion of glucose into cellular protein, for example, proceeds via precursor metabolites formed in the fuelling reactions, further via building blocks (in this case amino acids) formed in the biosynthetic reactions, and finally through polymerization of the building blocks (or amino acids). In the fuelling reactions there are many more intermediates than the precursor metabolites, and similarly there is a large number of intermediates in the conversion of precursor metabolites into building blocks. The number of cellular metabolites is therefore quite large, but still they only account for a small fraction of the total biomass (Table 4.1). The reason for this is the *en bloc* control of the individual reaction rates in the biosynthetic pathways mentioned above. Furthermore, the high affinity of enzymes for the reactants ensures that each metabolite can be maintained at a very low concentration even at a high flux through the pathway (see Box 4.2).

This control of the individual reactions in long pathways is very important for cell function, but it also means that in a

Box 4.2 **Control of metabolite levels in biochemical pathways**

The level of intracellular metabolites is normally very low. This is due to tight regulation of the enzyme levels and to the high affinity most enzymes have towards the reactants. To illustrate this consider two reactions of a pathway – one forming the metabolite X_i and the other consuming this metabolite:

$$\text{K } X_{i-1} \xrightarrow{v_i} X_i \xrightarrow{v_{i+1}} X_{i+1}\text{K}$$

Assuming that there is no allosteric regulation of the two enzyme-catalysed reactions the kinetics can be described with reversible Michaelis–Menten kinetics:

$$v_i = \frac{v_{i,\max}\left(\dfrac{c_{i-1}}{K_{i-1}} - \dfrac{c_i}{K_i}\right)}{1 + \dfrac{c_{i-1}}{K_{i-1}} + \dfrac{c_i}{K_i}}$$

where c_i is the metabolite concentration and $v_{i,\max}$ expresses the enzyme activity. If the rate of the first reaction increases drastically, e.g. due to an increase in the concentration of the metabolite X_{i-1}, the metabolite X_i will accumulate. This will lead to an increase in the second reaction rate and a decrease in the first reaction rate. Consequently the concentration of metabolite X_i will decrease again. The parameters K_i and K_{i-1} quantify the affinity of the enzyme for the reactant and the product in each reaction, and generally these are in the order of a few μM. Thus, even for low metabolite concentrations (in the order of 10 times K_i) the enzyme will be saturated and the reaction rate will be close to $v_{i,\max}$, but typically the metabolite concentration is of the order of K_i because hereby the enzyme can respond rapidly to changes in the metabolite level, and metabolite accumulations can be avoided.

quantitative description of cell growth it is not necessary to consider the kinetics of all the individual reactions, and this of course immediately results in a significant reduction in the degree of complexity. Consideration of the kinetics of individual enzymes or reactions is therefore necessary only when the aim of the study is to quantify the relative importance of a particular reaction in a pathway.

4.2.1 *Stoichiometry*

The first step in a quantitative description of cellular growth is to specify the stoichiometry for those reactions that are to be considered for analysis. For this purpose it is important to distinguish between substrates, metabolic products, intracellular metabolites and biomass constituents (Stephanopoulos *et al.*, 1998):

- A **substrate** is a compound present in the sterile medium and which can be further metabolized by or directly incorporated into the cell.
- A **metabolic product** is a compound produced by the cells and which is excreted to the extracellular medium.

- **Biomass constituents** are pools of macromolecules which make up the biomass, e.g. RNA, DNA, protein, lipids, carbohydrates, but also macromolecular products accumulating inside the cell, e.g. a polysaccharide or a non-secreted heterologous protein.
- **Intracellular metabolites** are all other compounds within the cell, i.e. glycolytic intermediates, precursor metabolites and building blocks.

Above there is a distinction between biomass constituents and intracellular metabolites, because the timescales of their turnovers in cellular reactions are very different, i.e. intracellular metabolites have a very fast turnover (typically in the range of seconds) compared with that of macromolecules (typically in the range of hours). This means that on the timescale of growth the intracellular metabolite pools can be assumed to be in pseudo steady state.

With the goal of specifying a general stoichiometry for biochemical reactions we consider a system where N substrates are converted to M metabolic products and Q biomass constituents. The conversions are carried out in J reactions in which K intracellular metabolites participate as pathway intermediates. The substrates are termed S_i, the metabolic products are termed P_i, the biomass constituents are termed $X_{macro,i}$, and the intracellular metabolites are termed $X_{met,i}$. With these definitions the general stoichiometry for the jth reaction can be specified as:

$$\sum_{i=1}^{N}\alpha_{ji}S_i + \sum_{i=1}^{M}\beta_{ji}P_i + \sum_{i=1}^{Q}\gamma_{ji}X_{macro,i} + \sum_{i=1}^{K}g_{ji}X_{met,i} = 0 \quad j = 1,\ldots,J$$

(4.3)

Here α_{ji} is a stoichiometric coefficient for the ith substrate, β_{ji} is a stoichiometric coefficient for the ith metabolic product, γ_{ji} is a stoichiometric coefficient for the ith macromolecular pool, and g_{ji} is a stoichiometric coefficient for the ith intracellular metabolite. All the stoichiometric coefficients are with sign. Thus, all compounds which are consumed in the jth reaction have negative stoichiometric coefficients whereas all compounds which are produced have positive stoichiometric coefficients. Furthermore, compounds which do not participate in the jth reaction have a stoichiometric coefficient of zero.

If there are many cellular reactions, i.e. J is large, it is convenient to write the stoichiometry for all the J cellular reactions in a compact form using matrix notation:

$$AS + BP + \Gamma X_{macro} + GX_{met} = 0 \qquad (4.4)$$

where the matrices **A**, **B**, Γ and **G** are stoichiometric matrices containing stoichiometric coefficients in the J reactions for the substrates, metabolic products, biomass constituents and pathway intermediates, respectively. In these matrices rows represents reactions and columns metabolites, i.e. the element in the jth row and the ith column of $\mathbf{A}$ specifies the stoichiometric coefficient for the ith substrate in the jth reaction. Formulation of the stoichiometry in matrix form may seem rather complex, but for large models it is much easier to handle, and especially if the model is to be used for analysis of pathways (e.g. metabolic flux analysis, see Chapter 7) the use of matrices is convenient. However, if the model is simple, i.e. only a few reactions, a few substrates, and a few metabolic products are considered, it is generally more convenient to use the more simple stoichiometric representation in equation (4.5).

4.2.2 Reaction rates

The stoichiometry of the individual reaction is the basis of any quantitative analysis. However, of equal importance is specification of the rate of the individual reactions. Normally the rate of a chemical reaction is given as the forward rate which if termed v_i specifies that a compound which has a stoichiometric coefficient β in the ith reaction is formed with the rate βv_i. Normally the stoichiometric coefficient for one of the compounds is arbitrarily set to 1, whereby the forward reaction rate becomes equal to the consumption or production of this compound in this particular reaction. For this reason the forward reaction rate is normally specified with the unit moles (or g) h^{-1}, or if the total amount of biomass is taken as reference (so-called specific rates) with the unit moles (or g) (g DW h)$^{-1}$.

For calculation of the overall production or consumption rate we have to sum the contributions from the different reactions, i.e. the total specific consumption rate of the ith substrate equals the sum of substrate consumptions in all the J reactions:

$$r_{s,i} = -\sum_{j=1}^{J}\alpha_{ji}v_j \qquad (4.5)$$

The stoichiometric coefficients for substrates are generally negative, i.e. the specific formation rate of the ith substrate in the jth reaction given by $\alpha_{ji}v_j$ is negative, but the specific substrate uptake rate is normally considered as positive, and a minus sign is therefore introduced in equation (4.5). For the specific formation rate of the ith metabolic product we have similarly:

$$r_{p,i} = \sum_{j=1}^{J}\beta_{ji}v_j \qquad (4.6)$$

Equations (4.5) and (4.6) specify some very important relations between what can be directly measured, namely the specific substrate uptake rates and the specific product formation rates, and the rates of the reactions in the metabolic model. If a compound is consumed or formed in only one reaction it is quite clear that we can get a direct measurement of this reaction rate – something that is the basis of metabolic flux analysis using metabolite balancing (see Chapter 7). For the biomass constituents and the intracellular metabolites we can specify similar expressions for the net formation rate in all the J reactions:

$$r_{macro,i} = \sum_{j=1}^{J} \gamma_{ji} v_j \tag{4.7}$$

$$r_{met,i} = \sum_{j=1}^{J} g_{ji} v_j \tag{4.8}$$

These rates are net specific formation rates, because a compound may be formed in one reaction and consumed in another, and the rates specify the net results of consumption and formation in all the J cellular reactions. Thus, if $r_{met,i}$ is positive there is a net formation of the ith intracellular metabolite and if it is negative there is a net consumption of this metabolite. Finally if $r_{met,i}$ is zero the rates of formation of the ith metabolite exactly balances its consumption.

If the forward reaction rates for the J cellular reactions are collected in the rate vector $\mathbf{v}$ the summations in equations (4.5) to (4.8) can be formulated in matrix notation as:

$$\mathbf{r}_s = -\mathbf{A}^T \mathbf{v} \tag{4.9}$$

$$\mathbf{r}_p = \mathbf{B}^T \mathbf{v} \tag{4.10}$$

$$\mathbf{r}_{macro} = \mathbf{\Gamma}^T \mathbf{v} \tag{4.11}$$

$$\mathbf{r}_{met} = \mathbf{G}^T \mathbf{v} \tag{4.12}$$

Here $\mathbf{r}_s$ is a rate vector containing the specific uptake rates of the N substrates, $\mathbf{r}_p$ a vector containing the specific formation rates of the M metabolic products, $\mathbf{r}_{macro}$ a vector containing the net specific formation rate of the Q biomass constituents, and $\mathbf{r}_{met}$ is a vector containing the net specific formation rate of the K intracellular metabolites. Notice that what appears in the matrix equations are the transposed stoichiometric matrices, which are formed from the stoichiometric matrices by converting columns into rows and vice versa (see Example 4a). Equation (4.7) gives the net specific formation rate of biomass constituents, and because the intracellular metabolites only represent a small fraction of the total biomass the specific growth rate μ of the total biomass is given as the sum of formation rates for all the macromolecular constituents:

$$\mu = \sum_{i=1}^{Q} r_{\text{macro},i} = \mathbf{1}_Q^T \mathbf{r}_{\text{macro}} = \mathbf{1}_Q^T \mathbf{\Gamma}^T \mathbf{v} \qquad (4.13)$$

where $\mathbf{1}_Q^T$ is a Q-dimensional row vector with all elements being 1. Equation (4.13) is very fundamental because it links the information supplied by a detailed metabolic model with the macroscopic (and measurable) parameter μ. It clearly specifies that the formation rate of biomass is represented by a sum of formation of many different biomass constituents (or macromolecular pools), a point that will be discussed further in section 4.4.3.

4.2.3 *Yield coefficients and linear rate equations*

The overall yield, e.g. how much carbon in the glucose ends up in the metabolite of interest, is a very important design parameter in many fermentation processes. This overall yield is normally represented in the form of so-called **yield coefficients,** which can be considered as relative rates (or fluxes) towards the product of interest with a certain compound as reference, often the carbon source or the biomass. These yield coefficients therefore have the units mass per unit mass of the reference, e.g. moles of penicillin formed per mole of glucose consumed or g protein formed per g biomass formed. A typical yield coefficient in the design and operation of aerobic fermentations is the respiratory quotient RQ, which specifies the moles of carbon dioxide formed per mole of oxygen consumed (see also Example 4a). Several different formulations of the yield coefficients can be found in the literature. Here we will use the formulation of Nielsen and Villadsen (1994) where the yield coefficient is stated with a double subscript Y_{ij} which states that a mass of j is formed or consumed per mass of i formed or consumed. With the ith substrate as the reference compound the yield coefficients are given by:

$$Y_{s_i s_j} = \frac{r_{s,j}}{r_{s,i}} \qquad (4.14)$$

$$Y_{s_i p_j} = \frac{r_{p,j}}{r_{s,i}} \qquad (4.15)$$

$$Y_{s_i x} = \frac{\mu}{r_{s,i}} \qquad (4.16)$$

In the classical description of cellular growth introduced by Monod (1942) (see section 4.4.2), the yield coefficient Y_{sx} was taken to be constant, and all the cellular reactions were lumped into a single overall growth reaction where substrate is converted to biomass. However, in 1959 it was shown by Herbert that the yield of biomass on substrate is not constant. In order to describe this he introduced the concept of **endogenous metabolism,** and specified substrate consumption for this process in addition to that for biomass synthesis. In the same year Luedeking and Piret found that lactic acid bacteria produce lactic acid at non-growth conditions, which was consistent with an endogenous metabolism of the cells. Their results indicated a linear correlation between the specific lactic acid production rate and the specific growth rate:

$$r_p = a\mu + b \tag{4.17}$$

In 1965 Pirt introduced a similar linear correlation between the specific substrate uptake rate and the specific growth rate, and he suggested use of the term **maintenance**, which now is the most commonly used term for the endogenous metabolism. The linear correlation of Pirt takes the form

$$r_s = Y_{xs}^{\text{true}}\,\mu + m_s \tag{4.18}$$

where Y_{xs}^{true} is referred to as the true yield coefficient and m_s as the maintenance coefficient. With the introduction of the linear correlations the yield coefficients can obviously not be constants. Thus for the biomass yield on the substrate:

$$Y_{sx} = \frac{\mu}{Y_{xs}^{\text{true}}\,\mu + m_s} \tag{4.19}$$

which shows that Y_{sx} decreases at low specific growth rates where an increasing fraction of the substrate is used to meet the maintenance requirements of the cell. For large specific growth rates the yield coefficient approaches the reciprocal of Y_{xs}^{true}, i.e. Y_{sx} becomes equal to Y_{sx}^{true}. A compilation of true yield coefficients and maintenance coefficients for various microbial species is given in Table 4.2.

Table 4.2 True yield and maintenance coefficients for different microbial species and growth on glucose or glycerol

Organism	Substrate	Y_{xs}^{true} (g (g DW)$^{-1}$)	m_s [g (g DW h)$^{-1}$]
Aspergillus awamori	Glucose	1.92	0.016
Aspergillus nidulans		1.67	0.020
Candida utilis		2.00	0.031
Escherichia coli		2.27	0.057
Klebsiella aerogenes		2.27	0.063
Penicillium chrysogenum		2.17	0.021
Saccharomyces cerevisiae		1.85	0.015
Aerobacter aerogenes	Glycerol	1.79	0.089
Bacillus megatarium		1.67	–
Klebsiella aerogenes		2.13	0.074

Data are taken from Nielsen and Villadsen (1994).

The empirically derived linear correlations are very useful for correlating growth data, especially in steady state continuous cultures where linear correlations similar to equation (4.18) were found for most of the important specific rates. The remarkable robustness and general validity of the linear correlations indicates that they have a fundamental basis, and this basis is likely to be the continuous supply and consumption of ATP, because these two processes are tightly coupled in all cells. Thus the role of the energy producing substrate is to provide ATP to drive both the biosynthetic and polymerization reactions of the cell and the different maintenance processes according to the linear relationship:

$$r_{ATP} = Y_{xATP}\mu + m_{ATP} \qquad (4.20)$$

which is a formal analogue to the linear correlation of Pirt. Equation (4.20) states that ATP produced balances the consumption for growth and for maintenance, and if the ATP yield on the energy producing substrate is constant, i.e. r_{ATP} is proportional to r_s, it is quite obvious that equation (4.20) can be used to derive the linear correlation in equation (4.18) as illustrated in Example 4a. Notice that Y_{xATP} in equation (4.20) actually is a true yield coefficient, but it is normally specified without the superscript 'true'.

The concept of balancing ATP production and consumption can be extended to other cofactors, e.g. NADH and NADPH, and hereby it is possible to derive linear rate equations for three different cases (Nielsen and Villadsen, 1994):

- anaerobic growth where ATP is supplied by substrate level phosphorylation
- aerobic growth without metabolite formation
- aerobic growth with metabolite formation

For aerobic growth with metabolite formation the specific substrate uptake rate takes the form:

$$r_s = Y_{xs}^{true}\mu + Y_{ps}^{true}r_p + m_s \qquad (4.21)$$

This linear rate equation can be interpreted as a metabolic model with three reactions:

- conversion of substrate to biomass with a stoichiometric coefficient Y_{xs}^{true} for the substrate and a forward reaction rate equal to the specific growth rate
- conversion of substrate to the metabolic product with a stoichiometric coefficient Y_{ps}^{true} for the substrate and a forward reaction rate equal to the specific product formation rate
- metabolism of substrate to meet the maintenance requirements (normally the substrate is oxidized to carbon dioxide) with the rate m_s.

Consequently the stoichiometry for these three reactions can be specified as:

$$-Y_{xs}^{true}S + X = 0 \; ; \; \mu \tag{4.22}$$

$$-Y_{ps}^{true}S + P = 0 \; ; \; r_p \tag{4.23}$$

$$-S = 0 \; ; \; m_s \tag{4.24}$$

With this stoichiometry the linear rate equation (4.21) can easily be derived using equation (4.5), i.e. the overall specific substrate consumption rate is the sum of substrate consumption for growth, metabolite formation and maintenance.

Thus, it is important to distinguish between true yield coefficients (which are rather stoichiometric coefficients) and overall yield coefficients, which can be taken to be stoichiometric coefficients in **one lumped reaction** (often referred to as the black box model, see section 4.2.4) which represents all the cellular processes, i.e.

$$-Y_{xs}S + Y_{xp}P + X = 0 \; ; \; \mu \tag{4.25}$$

Even though the true yield coefficients may be considered as true parameters for a given cellular system, they cannot be because they still represent overall stoichiometric coefficients in lumped reactions, e.g. reaction (4.22) is the sum of all reactions involved in the conversion of substrate into biomass. If, for example, the environmental conditions change, a different set of metabolic routes may be activated and this may result in a change in the overall recovery of carbon in each of the three processes mentioned above, i.e. the true yield coefficients change. Even the more fundamental $Y_{x\,ATP}$ cannot be taken to be constant, as illustrated in a detailed analysis of lactic acid bacteria (Benthin *et al.*, 1994).

Example 4a. Metabolic model for aerobic growth of *Saccharomyces cerevisiae*

To illustrate the derivation of the linear rate equations for an aerobic process with metabolite formation we consider a simple metabolic model for the yeast *Saccharomyces cerevisiae*. For this purpose we set up a stoichiometric model which summarizes the overall cellular metabolism, and based on assumptions of pseudo steady state for ATP, NADH and NADPH, linear rate equations can be derived where the specific uptake rates for glucose and oxygen and the specific carbon dioxide formation rate are given as functions of the specific growth rate. Furthermore, by evaluating the parameters in these linear rate equation expressions, which can be done from a comparison with experimental data, information on key energetic parameters may be extracted.

Table 4.3 Macromolecular composition of *Saccharomyces cerevisiae*

Macromolecule	Content $(g\ (g\ DW)^{-1})$
Protein	0.39
Polysaccharides + trehalose	0.39
DNA + RNA	0.11
Phospholipids	0.05
Triacylglycerols	0.02
Sterols	0.01
Ash	0.03

From an analysis of all the biosynthetic reactions the overall stoichiometry for synthesis of the constituents of a *S. cerevisiae* cell can be specified as (Oura, 1983):

$$CH_{1.62}O_{0.53}N_{0.15} + 0.12\ CO_2 + 0.397\ NADH - 1.12\ CH_2O - 0.15\ NH_3 - Y_{xATP}\ ATP - 0.212\ NADPH = 0 \tag{a1}$$

The stoichiometry (a1) holds for a cell with the composition specified in Table 4.3, and the substrate is glucose and inorganic salts, i.e. ammonia is the N source. The stoichiometry is given on a C-mole basis, i.e. glucose is specified as CH_2O, and the elemental composition of the biomass was calculated from the macromolecular composition to be $CH_{1.62}O_{0.53}N_{0.15}$ (see also Table 4.3). The ATP and NADPH required for biomass synthesis are supplied by the catabolic pathways, and excess NADH formed in the biosynthetic reactions is, together with NADH formed in the catabolic pathways, reoxidized by transfer of electrons to oxygen via the electron transport chain. Reactions (a2)–(a5) specify the overall stoichiometry for the catabolic pathways. Reaction (a2) specifies NADPH formation by the PP pathway, where glucose is completely oxidized to CO_2; reaction (a3) is the overall stoichiometry for the combined EMP pathway and the TCA cycle; reaction (a4) is the fermentative glucose metabolism where glucose is converted to ethanol (this reaction only runs at high glucose uptake rates); and finally reaction (a5) is the overall stoichiometry for the oxidative phosphorylation, where the P/O ratio is the overall (or operational) P/O ratio for the oxidative phosphorylation.

$$CO_2 + 2\ NADPH - CH_2O = 0 \tag{a2}$$

$$CO_2 + 2\ NADH + 0.667\ ATP - CH_2O = 0 \tag{a3}$$

$$CH_3O_{0.5} + 0.5\ CO_2 + 0.5\ ATP - 1.5\ CH_2O = 0 \tag{a4}$$

$$P/O\ ATP - 0.5\ O_2 - NADH = 0 \tag{a5}$$

Finally, consumption of ATP for maintenance is included simply as a reaction where ATP is used:

$$-ATP = 0 \tag{a6}$$

Note that with the stoichiometry given on a C-mole basis the stoichiometric coefficients extracted from the biochemistry, e.g. formation of 2 moles ATP per mole glucose in the EMP pathway, are divided by six, because glucose contains six C-moles per mole.

Above, the stoichiometry is written as in equation (4.3), but we can easily convert it to the more compact matrix notation of equation (4.4):

$$
\begin{pmatrix} -1.120 & 0 \\ -1 & 0 \\ -1 & 0 \\ -1.5 & 0 \\ 0 & -0.5 \\ 0 & 0 \end{pmatrix} \begin{pmatrix} S_{glc} \\ S_{o_2} \end{pmatrix} + \begin{pmatrix} 0 & 0.120 \\ 0 & 1 \\ 0 & 1 \\ 1 & 0.5 \\ 0 & 0 \\ 0 & 0 \end{pmatrix} \begin{pmatrix} P_{eth} \\ P_{co_2} \end{pmatrix} +
$$

$$
\begin{pmatrix} 1 \\ 0 \\ 0 \\ 0 \\ 0 \\ 0 \end{pmatrix} X + \begin{pmatrix} -Y_{xATP} & 0.397 & -0.212 \\ 0 & 0 & 2 \\ 0.667 & 2 & 0 \\ 0.5 & 0 & 0 \\ P/O & -1 & 0 \\ -1 & 0 & 0 \end{pmatrix} \begin{pmatrix} X_{ATP} \\ X_{NADH} \\ X_{NADPH} \end{pmatrix} = \begin{pmatrix} 0 \\ 0 \\ 0 \\ 0 \\ 0 \\ 0 \end{pmatrix} \tag{a7}
$$

where X represents the biomass.

We now collect the forward reaction rates for the six reactions in the rate vector v given by:

$$
v = \begin{pmatrix} \mu \\ v_{PP} \\ v_{EMP} \\ r_{eth} \\ v_{OP} \\ m_{ATP} \end{pmatrix}
$$

In analogy with equation (4.18) we balance the production and consumption of the three cofactors ATP, NADH and NADPH. This gives the three equations:

$$-Y_{xATP}\mu + 0.667v_{EMP} + 0.5r_{eth} + P/Ov_{OP} - m_{ATP} = 0 \tag{a8}$$

$$0.397\mu + 2v_{EMP} - v_{OP} = 0 \tag{a9}$$

$$-0.212\mu + 2v_{PP} = 0 \tag{a10}$$

Notice that these balances correspond to zero net specific formation rates for the three cofactors, and the three balances can therefore also be derived using equation (4.12):

$$r_{met} = G^T v = \begin{pmatrix} -Y_{xATP} & 0 & 0.667 & 0.5 & P/O & -1 \\ 0.397 & 0 & 2 & 0 & -1 & 0 \\ -0.212 & 2 & 0 & 0 & 0 & 0 \end{pmatrix} \begin{pmatrix} \mu \\ v_{PP} \\ v_{EMP} \\ r_{eth} \\ v_{OP} \\ m_{ATP} \end{pmatrix} = \begin{pmatrix} 0 \\ 0 \\ 0 \end{pmatrix}$$

(a11)

In addition to the three balances (a8)–(a10) we have the relationships between the reaction rates and the specific substrate uptake rates and the specific product formation rate given by equations (a5) and (a6), or using the matrix notation of equations (a9) and (a10):

$$\begin{pmatrix} r_{glc} \\ r_{O_2} \end{pmatrix} = -\begin{pmatrix} -1.120 & -1 & -1 & -1.5 & 0 & 0 \\ 0 & 0 & 0 & 0 & -0.5 & 0 \end{pmatrix} \begin{pmatrix} \mu \\ v_{PP} \\ v_{EMP} \\ r_{eth} \\ v_{OP} \\ m_{ATP} \end{pmatrix}$$

(a12)

$$= \begin{pmatrix} 1.120\mu + v_{PP} + v_{EMP} + 1.5r_{eth} \\ 0.5v_{OP} \end{pmatrix}$$

$$\begin{pmatrix} r_{eth} \\ r_{CO_2} \end{pmatrix} = -\begin{pmatrix} 0 & 0 & 0 & 1 & 0 & 0 \\ 0.120 & 1 & 1 & 0.5 & 0 & 0 \end{pmatrix} \begin{pmatrix} \mu \\ v_{PP} \\ v_{EMP} \\ r_{eth} \\ v_{OP} \\ m_{ATP} \end{pmatrix}$$

(a13)

$$= \begin{pmatrix} r_{eth} \\ 1.120\mu + v_{PP} + v_{EMP} + 0.5r_{eth} \end{pmatrix}$$

Clearly the specific ethanol production rate is equal to the rate of reaction (a5) because the stoichiometric coefficient for ethanol in this reaction is 1 and it is the only reaction where ethanol is involved. Using the combined set of equations (a11)–(a13) the four reaction rates v_{EMP}, v_{PP}, v_{OP} and m_{ATP} can be eliminated and the linear rate equations (a14)–(a16) can be derived:

$$r_{glc} = (a + 1.226)\mu + (1.5 - b)r_{eth} + c$$
$$= Y_{xs}^{true}\mu + Y_{ps}^{true}r_{eth} + m_s$$

(a14)

$$r_{CO_2} = (a + 0.226)\mu + (0.5 - b)r_{eth} + c$$
$$= Y_{xc}^{true}\mu + Y_{pc}^{true}r_{eth} + m_c \tag{a15}$$

$$r_{O_2} = (a + 0.229)\mu - br_{eth} + c = Y_{xo}^{true}\mu + Y_{po}^{true}r_{eth} + m_o \tag{a16}$$

The three common parameters a, b and c are functions of the energetic parameters Y_{xATP}, m_{ATP} and the P/O ratio according to equations (a17)–(a19):

$$a = \frac{Y_{xATP} - 0.458P/O}{0.667 + 2P/O} \tag{a17}$$

$$b = \frac{0.5}{0.667 + 2P/O} \tag{a18}$$

$$c = \frac{m_{ATP}}{0.667 + 2P/O} \tag{a19}$$

If there is no ethanol formation, which is the case at low specific glucose uptake rates, equation (a14) reduces to the linear rate equation (4.18) introduced by Pirt, but the parameters of the correlation are determined by basic energetic parameters of the cells. It is seen that the parameters in the linear correlations are coupled via the balances for ATP, NADH and NADPH, and the three true yield coefficients cannot take any value. Furthermore, the maintenance coefficients are the same. This is due to the use of the units C-moles per C-mole biomass per h for the specific rates. If other units are used for the specific rates the maintenance coefficients will not take the same values, but they would still be proportional. This coupling of the parameters shows that there are only three degrees of freedom in the system, and one actually only has to determine two yield coefficients and one maintenance coefficient – the other parameters can be calculated using the three equations (a14)–(a16).

The derived linear rate equations are certainly useful for correlating experimental data, but they also allow evaluation of the key energetic parameters Y_{xATP}, m_{ATP} and the operational P/O ratio. Thus if the true yield coefficients and the maintenance coefficients of equations (a14)–(a16) are estimated, the values of a, b and c can be found and these three parameters relate the three energetic parameters through equations (a17)–(a19). Thus from one of the ethanol yield coefficients b can be found, and thereafter the P/O ratio can be found. Then m_{ATP} can be found from one of the main-tenance coefficients, and finally Y_{xATP} can be found from one of the biomass yield coefficients. In practice it is, however, difficult to extract sufficiently precise values of the true yield coefficients from experimental data to estimate the energetic parameters – especially since the three parameters a, b and c are closely correlated (especially b and c). However, if either the P/O ratio or Y_{xATP} is known, equations (a14)–(a16) allow an estimation of the two remaining unknown energetic parameters. Consider the situation where there is no ethanol formation. Here the true yield coefficient for biomass is 1.48 C-moles glucose (C-mole biomass)$^{-1}$ and the maintenance coefficient (equal to b) is 0.012 C-moles glucose (C-mole biomass h)$^{-1}$ (both values taken from Table 4.2). Thus a is equal to 0.254 moles ATP (C-mole biomass)$^{-1}$. If the operational P/O ratio is about 1.5 (which is a reasonable value for *S. cerevisiae*) we find that Y_{xATP} is 1.62 moles ATP (C-mole biomass)$^{-1}$ or about 67 mmoles ATP (g DW)$^{-1}$. Similarly we find m_{ATP} to be about 2 mmoles (g DW h)$^{-1}$.

In connection with baker's yeast production it is important to maximize the yield of biomass on glucose:

$$Y_{sx} = \frac{\mu}{(a + 1.226)\mu + (1.5 - b)r_{eth} + c} \tag{a20}$$

Clearly this can best be done if ethanol production is avoided. Thus, the glucose uptake rate is to be controlled below a level where there is respiro-fermentative metabolism. A very good indication of whether there is respiro-fermentative metabolism is the respiratory quotient RQ:

$$RQ = \frac{(a + 0.226)\mu + (0.5 - b)r_{eth} + c}{(a + 0.229)\mu - br_{eth} + c} \tag{a21}$$

If there is no ethanol production RQ will clearly be close to 1 (independent of the specific growth rate), whereas if there is ethanol production RQ will be above 1 and will increase with r_{eth} (b is always less than 0.5). From measurements of carbon dioxide and oxygen in the exhaust gas the RQ can be evaluated and if it is above 1 there is respiro-fermentative metabolism resulting in a low yield of biomass on sugar. Clearly the cells are exposed to too high a sugar concentration, and the feed to the reactor should be reduced (typically the baker's yeast production is operated as a fed-batch process, see section 3.4).

4.2.4 The black box model

In the black box model of cellular growth all the cellular reactions are lumped into a single reaction. In this overall reaction the stoichiometric coefficients are identical to the yield coefficients (see also equation (4.25)), and it can therefore be presented as:

$$X + \sum_{i=1}^{M} Y_{xp_i} P_i - \sum_{i=1}^{N} Y_{xs_i} S_i = 0 \tag{4.26}$$

Because the stoichiometric coefficient for biomass is 1, the forward reaction rate is given by the specific growth rate of the biomass, which together with the yield coefficients completely specify the system. As discussed in the previous section the yield coefficients are not constants, and the black box model can therefore not be applied to correlate, for instance, the specific substrate uptake rate with the specific growth rate. However, it is very useful for validation of experimental data because it can form the basis for setting up elemental balances. Thus, in the black box model there are $(M + N + 1)$ parameters: M yield coefficients for the metabolic products, N yield coefficients for the substrates and the forward reaction rate μ. Because mass is conserved in the overall conversion of substrates to metabolic products and biomass, the $(M + N + 1)$ parameters of the black box model are not completely independent but must satisfy several constraints. Thus, the elements flowing into the system must balance the elements flowing out of the system, e.g. the carbon entering the system via the substrates has to be recovered in the metabolic products and biomass. Each element considered in the black box obviously yields one constraint. Thus a carbon balance gives:

Table 4.4 Elemental composition of biomass for several microorganisms

Microorganism	Elemental composition	Ash content (w/w %)	Conditions
Candida utilis	$CH_{1.83}O_{0.46}N_{0.19}$	7.0	Glucose limited, $D = 0.05$ h^{-1}
	$CH_{1.87}O_{0.56}N_{0.20}$	7.0	Glucose limited, $D = 0.45$ h^{-1}
	$CH_{1.83}O_{0.54}N_{0.10}$	7.0	Ammonia limited, $D = 0.05$ h^{-1}
	$CH_{1.87}O_{0.56}N_{0.20}$	7.0	Ammonia limited, $D = 0.45$ h^{-1}
Klebsiella aerogenes	$CH_{1.75}O_{0.43}N_{0.22}$	3.6	Glycerol limited, $D = 0.10$ h^{-1}
	$CH_{1.73}O_{0.43}N_{0.24}$	3.6	Glycerol limited, $D = 0.85$ h^{-1}
	$CH_{1.75}O_{0.47}N_{0.17}$	3.6	Ammonia limited, $D = 0.10$ h^{-1}
	$CH_{1.73}O_{0.43}N_{0.24}$	3.6	Ammonia limited, $D = 0.80$ h^{-1}
Saccharomyces cerevisiae	$CH_{1.82}O_{0.58}N_{0.16}$	7.3	Glucose limited, $D = 0.080$ h^{-1}
	$CH_{1.78}O_{0.60}N_{0.19}$	9.7	Glucose limited, $D = 0.255$ h^{-1}
	$CH_{1.94}O_{0.52}N_{0.25}$	5.5	Unlimited growth
Escherichia coli	$CH_{1.77}O_{0.49}N_{0.24}$	5.5	Unlimited growth
	$CH_{1.83}O_{0.50}N_{0.22}$	5.5	Unlimited growth
	$CH_{1.96}O_{0.55}N_{0.25}$	5.5	Unlimited growth
	$CH_{1.93}O_{0.55}N_{0.25}$	5.5	Unlimited growth
Pseudomonas fluorescens	$CH_{1.83}O_{0.55}N_{0.26}$	5.5	Unlimited growth
Aerobacter aerogenes	$CH_{1.64}O_{0.52}N_{0.16}$	7.9	Unlimited growth
Penicillium chrysogenum	$CH_{1.70}O_{0.58}N_{0.15}$		Glucose limited, $D = 0.038$ h^{-1}
	$CH_{1.68}O_{0.53}N_{0.17}$		Glucose limited, $D = 0.098$ h^{-1}
Aspergillus niger	$CH_{1.72}O_{0.55}N_{0.17}$	7.5	Unlimited growth
Average	$CH_{1.81}O_{0.52}N_{0.21}$	6.0	

Compositions for *P. chrysogenum* are taken from Christensen *et al.* (1995); other data are taken from Roels (1983).

$$1 + \sum_{i=1}^{M} f_{p,i} Y_{xp_i} - \sum_{i=1}^{N} f_{s,i} Y_{xs_i} = 0 \qquad (4.27)$$

where $f_{s,i}$ and $f_{p,i}$ represent the carbon content (C-moles mole^{-1}) in the ith substrate and the ith metabolic product, respectively. In the above equation, the elemental composition of biomass is normalized with respect to carbon, i.e. it is represented by the form $CH_aO_bN_c$ (see also Example 4a). The elemental composition of biomass depends on its macromolecular content and, therefore, on the growth conditions and the specific growth rate (e.g. the nitrogen content is much lower under nitrogen limited conditions than under carbon limited conditions; see Table 4.4). However, except for extreme situations, it is reasonable to use the general composition formula $CH_{1.8}O_{0.5}N_{0.2}$ whenever the biomass composition is not exactly known. Often the elemental composition of substrates and metabolic products is normalized with respect to their carbon content, e.g. glucose is specified as CH_2O (see also Example 4a). Equation (4.27) is then written on a per C-mole basis as:

$$1 + \sum_{i=1}^{M} Y_{xp_i} - \sum_{i=1}^{N} Y_{xs_i} = 0 \qquad (4.28)$$

In equation (4.28) the yield coefficients have units of C-moles per C-mole biomass. Conversion to this unit from other units is illustrated in Box 4.3. Equation (4.28) is very useful for checking the consistency of experimental data. Thus, if the sum of carbon in the biomass and the metabolic products does not equal

---

Box 4.3 **Calculation of yields with respect to C-mole basis**

Yield coefficients are typically described as moles (g DW)$^{-1}$ or g (g DW)$^{-1}$. To convert the yield coefficients to a C-mole basis, information on the elemental composition and the ash content of biomass is needed. To illustrate the conversion, we calculate the yield of 0.5 g DW biomass (g glucose)$^{-1}$ on a C-mole basis. First, we convert the g DW biomass to an ash-free basis, i.e. determine the amount of biomass that is made up of carbon, nitrogen, oxygen and hydrogen (and, in some cases, also phosphorus and sulphur). With an ash content of 8% we have 0.92 g ash-free biomass (g DW biomass)$^{-1}$, which gives a

yield of 0.46 g ash-free biomass (g glucose)$^{-1}$. This yield can now be directly converted to a C-mole basis using the molecular weights in g C-mole^{-1} for ash-free biomass and glucose. With the standard elemental composition for biomass of $CH_{1.8}O_{0.5}N_{0.2}$ we have a molecular weight of 24.6 g ash-free biomass C-mole^{-1}, and therefore find a yield of $0.46/24.6 = 0.0187$ C-moles biomass (g glucose)$^{-1}$. Finally, by multiplication with the molecular weight of glucose on a C-mole basis (30 g C-mole^{-1}), a yield of 0.56 C-moles biomass (C-mole glucose)$^{-1}$ is found.

the sum of carbon in the substrates there is an inconsistency in the experimental data.

Similar to equation (4.27), balances can be written for all other elements participating in the conversion (4.26). Thus, the hydrogen balance will read:

$$a_x + \sum_{i=1}^{M} a_{p,i} Y_{xp_i} - \sum_{i=1}^{N} a_{s,i} Y_{xs_i} = 0 \qquad (4.29)$$

where $a_{s,i}$, $a_{p,i}$ and a_x represent the hydrogen content (moles C-mole^{-1} if a C-mole basis is used) in the ith substrate, the ith metabolic product and the biomass, respectively. Similarly we have for the oxygen and nitrogen balances:

$$b_x + \sum_{i=1}^{M} b_{p,i} Y_{xp_i} - \sum_{i=1}^{N} b_{s,i} Y_{xs_i} = 0 \qquad (4.30)$$

$$c_x + \sum_{i=1}^{M} c_{p,i} Y_{xp_i} - \sum_{i=1}^{N} c_{s,i} Y_{xs_i} = 0 \qquad (4.31)$$

where $b_{s,i}$, $b_{p,i}$ and b_x represent the oxygen content (moles C-mole^{-1}) in the ith substrate, the ith metabolic product and biomass, respectively, and $c_{s,i}$, $c_{p,i}$ and c_x represent the nitrogen content (moles C-mole^{-1}) in the ith substrate, the ith metabolic product and the biomass, respectively. Normally only these four balances are considered; balances for phosphate and sulphate may also be set up, but generally these elements are of minor importance. The four elemental balances (4.28) to (4.31) can be conveniently written by collecting the elemental composition of biomass, substrates and metabolic products in the columns of a matrix E, where the first column contains the elemental composition of biomass, columns 2 through $M + 1$ contain the elemental composition of the M metabolic products, and columns $M + 2$ to $M + N + 1$ contain the elemental composition of the N substrates. With the introduction of this matrix the four elemental balances can be expressed as:

$$\mathbf{E\,Y} = 0 \qquad (4.32)$$

where $\mathbf{Y}$ is a vector containing the yield coefficients (the substrate yield coefficients are given with a minus sign). With $N + M + 1$ variables, $N + M$ yield coefficients and the forward rate of reaction (4.26) and four constraints, the degree of freedom is $F = M + N + 1 - 4$. If exactly F variables are measured it may be possible to calculate the other rates by using the four algebraic equations given by (4.32), but, in this case, there are no redundancies left to check the consistency of the data. For this reason, it is advisable to strive for more measurements than the degrees of freedom of the system.

Example 4b. Elemental balances in a simple black box model
Consider the aerobic cultivation of the yeast *Saccharomyces cerevisiae* on a defined, minimal medium, i.e. glucose is the carbon and energy source and ammonia is the nitrogen source. During aerobic growth, the yeast oxidizes glucose completely to carbon dioxide. However, at very high glycolytic fluxes, a bottleneck in the oxidation of pyruvate leads to ethanol formation. Thus, at high glycolytic fluxes, both ethanol and carbon dioxide should be considered as metabolic products. Finally, water is formed in the cellular pathways and this is also included as a product in the overall reaction. Thus, the black box model for this system is:

$$X + Y_{xe} \text{ ethanol} + Y_{xc}\, CO_2 + Y_{xw}\, H_2O - Y_{xs} \text{ glucose} - Y_{xo}\, O_2 - Y_{xN}\, NH_3 = 0 \tag{b1}$$

which can be represented with the yield coefficient vector:

$$\mathbf{Y} = (1 \quad Y_{xe} \quad Y_{xc} \quad Y_{xw} \quad -Y_{xs} \quad -Y_{xo} \quad -Y_{xN})^T \tag{b2}$$

We now rewrite the conversion using the elemental composition of the substrates and metabolic products. For biomass we use the elemental composition of $CH_{1.83}O_{0.56}N_{0.17}$, and therefore have:

$$CH_{1.83}O_{0.56}N_{0.17} + Y_{xe}\, CH_3O_{0.5} + Y_{xc}\, CO_2 + Y_{xw}\, H_2O - Y_{xs}\, CH_2O - Y_{xo}\, O_2 - Y_{xN}\, NH_3 = 0 \tag{b3}$$

Some may find it difficult to identify $CH_3O_{0.5}$ as ethanol, but the advantage of using the C-mole basis becomes apparent immediately when we look at the carbon balance:

$$1 + Y_{xe} + Y_{xc} - Y_{xs} = 0 \tag{b4}$$

This simple equation is very useful for checking the consistency of experimental data. Thus, using the classical data of von Meyenburg (1969), we find $Y_{xe} = 0.713$, $Y_{xc} = 1.313$ and $Y_{xs} = 3.636$ at a dilution rate of $D = 0.3\ h^{-1}$ in a glucose-limited continuous culture. Obviously the data are not consistent as the carbon balance does not close. A more elaborate data analysis (Nielsen and Villadsen, 1994) suggests that the missing carbon is ethanol which could have evaporated as a result of ethanol stripping due to intensive aeration of the bioreactor.

Similarly using equation (4.31) we find that a nitrogen balance gives:

$$Y_x N = 0.17 \tag{b5}$$

If the yield coefficient for ammonia uptake and biomass formation do not conform with equation (b5), an inconsistency is identified in one of these two measurements or the nitrogen content of the biomass is different from that specified.

We now write all four elemental balances in terms of the matrix equation (4.32):

$$
E = \begin{pmatrix}
1 & 1 & 1 & 0 & 1 & 0 & 0 \\
1.83 & 3 & 0 & 2 & 2 & 0 & 3 \\
0.56 & 0.5 & 2 & 1 & 1 & 2 & 0 \\
0.17 & 0 & 0 & 0 & 0 & 0 & 1
\end{pmatrix}
\begin{matrix}
\leftarrow \text{carbon} \\
\leftarrow \text{hydrogen} \\
\leftarrow \text{oxygen} \\
\leftarrow \text{nitrogen}
\end{matrix}
\qquad (b6)
$$

where the rows indicate, respectively, the content of carbon, hydrogen, oxygen and nitrogen and the columns give the elemental composition of biomass, ethanol, carbon dioxide, water, glucose, oxygen and ammonia, respectively. Using equation (4.32) we find:

$$
\begin{pmatrix}
1 & 1 & 1 & 0 & 1 & 0 & 0 \\
1.83 & 3 & 0 & 2 & 2 & 0 & 3 \\
0.56 & 0.5 & 2 & 1 & 1 & 2 & 0 \\
0.17 & 0 & 0 & 0 & 0 & 0 & 1
\end{pmatrix}
\begin{pmatrix}
1 \\
Y_{xe} \\
Y_{xc} \\
Y_{xw} \\
-Y_{xs} \\
-Y_{xo} \\
-Y_{xN}
\end{pmatrix}
\qquad (b7)
$$

$$
= \begin{pmatrix}
1 + Y_{xe} + Y_{xc} - Y_{xs} \\
1.83 + 3Y_{xe} + 2Y_{xw} - 2Y_{xs} - 3Y_{xN} \\
0.56 + 0.5Y_{xe} + 2Y_{xc} + Y_{xw} - Y_{xs} - 2Y_{xo} \\
0.17 - Y_{xN}
\end{pmatrix} = \begin{pmatrix}
0 \\
0 \\
0 \\
0
\end{pmatrix}
$$

The first and last rows are identical to the balances derived in equations (b4) and (b5) for carbon and nitrogen, respectively. The balances for hydrogen and oxygen introduce two additional constraints. However, because the rate of water formation is impossible to measure, one of these equations must be used to calculate this rate (or yield). This leaves only one additional constraint from these two balances.

4.3   Mass balances for bioreactors

In the previous section we derived equations that relate the rates of the intracellular reaction with the rates of substrate uptake, metabolic product formation and biomass formation. These rates are the key elements in the dynamic mass balances for the substrates, the metabolic products and the biomass, which describe the change in time of the concentration of these state variables in a bioreactor. The bioreactor may be any type of device, ranging from a test tube or a shake flask to a well instrumented bioreactor. Figure 4.2 is a general representation

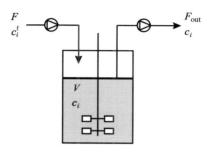

Figure 4.2

Bioreactor with addition of fresh, sterile medium and removal of spent medium. c_i^f is the concentration of the ith compound in the feed and c_i is the concentration of the ith compound in the spent medium. The bioreactor is assumed to be very well mixed (or ideal), so that the concentration of each compound in the spent medium becomes identical to its concentration in the bioreactor.

of a bioreactor. It has a volume $V(l)$ and it is fed with a stream of fresh, sterile medium with a flow rate $F(l\,h^{-1})$. Spent medium is removed with a flow rate of $F_{out}(l\,h^{-1})$. The medium in the bioreactor is assumed to be completely (or ideally) mixed, i.e. there is no spatial variation in the concentration of the different medium compounds. For small volume bioreactors ($< 1\,l$) (including shake flasks) this can generally be achieved through aeration and some agitation, whereas for laboratory, stirred tank bioreactors (1–$10\,l$) special designs may have to be introduced in order to ensure a homogenous medium (Sonnleitner and Fiechter, 1988; Nielsen and Villadsen, 1993). The bioreactor may be operated in many different modes of which we will only consider the three most common:

- **Batch**, where $F = F_{out} = 0$, i.e. the volume is constant.
- **Continuous**, where $F = F_{out} \neq 0$, i.e. the volume is constant.
- **Fed-batch** (or semi-batch), where $F \neq 0$ and $F_{out} = 0$, i.e. the volume increases.

These three different modes of reactor operation are discussed in the following but first we derive general dynamic mass balances for the substrates, metabolic products, biomass constituents and the biomass.

4.3.1 *Dynamic mass balances*

The basis for derivation of the general dynamic mass balances is the mass balance equation:

$$\text{Accumulated} = \text{Net formation rate} + \text{In} - \text{Out} \tag{4.33}$$

where the first term on the RHS is given by equations (4.5) to (4.8) for substrate, metabolic product, biomass constituents and intracellular metabolites, respectively. The term 'In' represents the flow of the compound into the bioreactor and the term 'Out' the flow of the compound out from the bioreactor. In the following we consider substrates, metabolic products, biomass constituents, intracellular metabolites and the total biomass separately.

We consider the ith substrate which is added to the bioreactor via the feed and is consumed by the cells present in the bioreactor. The mass balance for this compound is:

$$\frac{d(c_{s,i}V)}{dt} = -r_{s,i}xV + Fc_{s,i}^f - F_{out}c_{s,i} \tag{4.34}$$

where r_i is the specific consumption rate of the compound (moles (g DW h)$^{-1}$), $c_{s,i}$ is the concentration in the bioreactor, which is assumed to be the same as the concentration in the outlet (moles l^{-1}), $c_{s,i}^f$ is the concentration in the feed (moles l^{-1}), and x is the biomass concentration in the bioreactor (g DW l^{-1}). The first term in equation (4.34) is the accumulation term, the second term is the consumption (or reaction term), the third term is accounting for the inlet, and the last term is accounting for the outlet. Rearrangement of this equation gives:

$$\frac{dc_{s,i}}{dt} = -r_{s,i}x + \frac{F}{V}c_{s,i}^f - \left(\frac{F_{out}}{V} + \frac{1}{V}\frac{dV}{dt}\right)c_{s,i} \tag{4.35}$$

For a fed-batch reactor:

$$F = \frac{dV}{dt} \tag{4.36}$$

and $F_{out} = 0$ and so the term within the parentheses becomes equal to the so-called **dilution rate** given by:

$$D = \frac{F}{V} \tag{4.37}$$

For a continuous and a batch reactor the volume is constant, i.e. $dV/dt = 0$, and $F = F_{out}$, and so for these bioreactor modes also the term within the parentheses becomes equal to the dilution rate. Equation (4.35) therefore reduces to the mass balance (4.38) for any type of operation:

$$\frac{dc_{s,i}}{dt} = -r_{s,i}x + D(c_{s,i}^f - c_{s,i}) \tag{4.38}$$

The first term on the right-hand side of equation (4.38) is the volumetric rate of substrate consumption, which is given as the product of the specific rate of substrate consumption and the biomass concentration. The second term accounts for the addition and removal of substrate from the bioreactor. The term on the left-hand side of equation (4.38) is the accumulation term, which accounts for the change in time of the substrate, which in a batch reactor (where $D = 0$) equals the volumetric rate of substrate consumption.

Dynamic mass balances for the metabolic products are derived in analogy with those for the substrates and takes the form:

$$\frac{dc_{p,i}}{dt} = -r_{p,i}x + D(c_{p,i}^f - c_{p,i}) \tag{4.39}$$

where the first term on the right-hand side is the volumetric formation rate of the ith metabolic product. Normally the metabolic products are not present in the sterile feed to the bioreactor and $c_{p,i}^f$ is therefore often zero. In these cases the volumetric rate of product formation in a steady-state continuous reactor is equal to the dilution rate multiplied by the concentration of the metabolic product in the bioreactor (equal to that in the outlet).

With sterile feed the mass balance for the total biomass is derived directly:

$$\frac{dx}{dt} = (\mu - D)x \tag{4.40}$$

where $\mu(h^{-1})$ is the specific growth rate of the biomass given by equation (4.13).

For the biomass constituents we normally use the biomass as the reference, i.e. their concentrations are given with the biomass as the basis. In this case the mass balance for the ith biomass constituent is derived from (sterile feed is assumed):

$$\frac{d(X_{macro,i}xV)}{dt} = r_{macro,i}xV - F_{out}X_{macro,i}x \tag{4.41}$$

where $X_{macro,i}x$ is the concentration of the ith biomass component in the bioreactor (g l^{-1}) and $r_{macro,i}$ is the specific, net rate of formation of the ith biomass constituent. Rearrangement of equation (4.41) gives:

$$\frac{dX_{macro,i}}{dt} = r_{macro,i} - \left(\frac{F_{out}}{V} + \frac{1}{x}\frac{dx}{dt} + \frac{1}{V}\frac{dV}{dt}\right)X_{macro,i} \tag{4.42}$$

Again we have that for any mode of bioreactor operation:

$$D = \frac{F_{out}}{V} + \frac{1}{V}\frac{dV}{dt} \tag{4.43}$$

which together with the mass balance (4.40) for the total biomass concentration gives the mass balance

$$\frac{dX_{macro,i}}{dt} = r_{macro,i} - \mu X_{macro,i} \qquad (4.44)$$

where $X_{macro,i}$ is the concentration of the ith biomass constituent. Different units may be applied for the concentrations of the biomass constituents, but they are normally given as g (g DW)$^{-1}$, because then the sum of all the concentrations equals 1, i.e.

$$\sum_{i=1}^{Q} X_{macro,i} = 1 \qquad (4.45)$$

Furthermore, this unit corresponds with the experimentally determined macromolecular composition of cells where weight fractions are generally used. In equation (4.44) it is observed that the mass balance for the biomass constituents is completely independent of the mode of operation of the bioreactor, i.e. the dilution rate does not appear in the mass balance. However, there is indirectly a coupling via the last term, which accounts for dilution of the biomass constituents when the biomass expands due to growth. Thus if there is no net synthesis of a macromolecular pool, but the biomass still grows, the intracellular level decreases.

For intracellular metabolites it is not convenient to use the same unit for their concentrations as for the biomass constituents. These metabolites are dissolved in the matrix of the cell and it is therefore more appropriate to use the unit moles per liquid cell volume for the concentrations. The intracellular concentration can then be compared directly with the affinities of enzymes, typically quantified by their K_m values, which are normally given with the unit moles l^{-1}. If the concentration is known in one unit it is, however, easily converted to another unit if the density of the biomass (in the range of 1 g cell per ml cell) and the water content (in the range of 0.67 ml water per ml cell) are known. Even though a different unit is applied the biomass is still the basis and the mass balance for the intracellular metabolites therefore takes the same form:

$$\frac{dX_{met,i}}{dt} = r_{met,i} - \mu X_{met,i} \qquad (4.46)$$

where $X_{met,i}$ is the concentration of the ith intracellular metabolite. It is important to distinguish between concentrations of intracellular metabolites given in moles per liquid reactor volume and in moles per liquid cell volume. If concentrations are given in the former unit the mass balance will be completely different.

4.3.2   The batch reactor

This is the classical operation of the bioreactor, and it is used by many life scientists, because it can be carried out in a relatively simple experimental set-up. Batch experiments have the advantage of being easy to perform and by using shake flasks a large number of parallel experiments can be carried out. The disadvantage is that the experimental data are difficult to interpret because there are dynamic conditions throughout the experiment, i.e. the environmental conditions experienced by the cells vary with time. However, by using well instrumented bioreactors at least some variables, e.g. pH and dissolved oxygen tension, may be kept constant.

As mentioned in the previous section the dilution rate is zero for a batch reactor and the mass balances for the biomass and the limiting substrate therefore take the form:

$$\frac{dx}{dt} = \mu x \quad x(t = 0) = x_0 \tag{4.47}$$

$$\frac{dc_s}{dt} = -r_s x \quad c_s(t = 0) = c_{s,0} \tag{4.48}$$

where x_0 indicates the initial biomass concentration, which is obtained immediately after inoculation, and $c_{s,0}$ is the initial substrate concentration. According to the mass balance the biomass concentration will increase as indicated in Figure 4.3 and the substrate concentration will decrease until its concentration

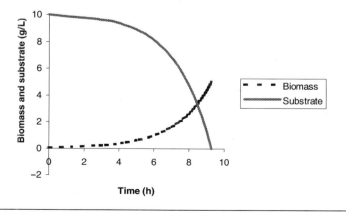

Figure 4.3
Batch fermentation described with Monod kinetics. The biomass concentration is found using equation (c2) and the corresponding substrate concentration is found from equation (c1). μ_{max} is 0.5 h^{-1}, K_s is 50 mg l^{-1} (a quite high value), and Y_{sx} is 0.5 g g^{-1}. The initial substrate concentration $c_{s,0}$ is 10 g l^{-1}. The substrate concentration decreases from 0.5 g l^{-1} to zero in less than 5 min, and this is the interesting substrate concentration range for estimation of K_s.

reaches zero and growth stops. Because the substrate concentration is zero at the end of the cultivation the overall yield of biomass on the substrate can be found from:

$$Y_{sx}^{overall} = \frac{x_{final} - x_0}{c_{s,0}} \qquad (4.49)$$

where x_{final} is the biomass concentration at the end of the cultivation. Normally $x_0 \ll x_{final}$, and the overall yield coefficient can therefore be estimated from the final biomass concentration and the initial substrate concentration alone. Notice that the yield coefficient determined from the batch experiment is the overall yield coefficient and not Y_{sx} or Y_{sx}^{true}. The yield coefficient Y_{sx} may well be time dependent as it is the ratio between the specific growth rate and the substrate uptake rate (see equation (4.16)). However, if there is little variation in these rates during the batch culture (e.g. if there is a long exponential growth phase and only a very short declining growth phase) the overall yield coefficient may be very similar to the yield coefficient. The true yield coefficient, on the other hand, is difficult to determine from a batch cultivation, because it requires information about the maintenance coefficients, which can hardly be determined from a batch experiment. However, in a batch cultivation the specific growth rate is close to its maximum throughout most of the growth phase, and the substrate consumption due to maintenance is therefore negligible, and according to equation (4.17) the true yield coefficient is close to the observed yield coefficient determined from the final biomass concentration.

4.3.3 The chemostat

A typical operation of the continuous bioreactor is the so-called **chemostat**, where the added medium is designed such that there is a **single limiting substrate**. This allows for controlled variation in the specific growth rate of the biomass. The advantage of the continuous bioreactor is that a steady state can be achieved, which allows for precise experimental determination of specific rates. Furthermore, by varying the feed flow rate to the bioreactor the environmental conditions can be varied, and valuable information concerning the influence of the environmental conditions on the cellular physiology can be obtained. The continuous bioreactor is attractive for industrial applications because the productivity can be high. However, often the titre, i.e. the product concentration, is lower than can be obtained in the fed-batch reactor, and it is therefore a trade-off between productivity and titre. Furthermore, it is rarely used in industrial processes because it is sensitive to contamination, e.g. via the feed stream, and to the appearance of spontaneously formed mutants that may out-compete the production strain. The disadvantage of the continuous bioreactor is that it is laborious to operate because large amounts of fresh, sterile medium have to be prepared. Other examples of continuous operation besides the chemostat are the **pH-stat**, where the feed flow is adjusted to maintain the pH constant in the bioreactor, and the **turbidostat**, where the feed flow is adjusted to maintain the biomass concentration at a constant level.

From the biomass mass balance (4.40) it can easily be seen that in a steady-state continuous reactor the specific growth rate equals the dilution rate:

$$\mu = D \tag{4.50}$$

Thus by varying the dilution rate (or the feed flow rate) in a continuous culture, different specific growth rates can be obtained. This allows detailed physiological studies of the cells when they are grown at a predetermined specific growth rate (corresponding to a certain environment experienced by the cells). At steady state the substrate mass balance (4.38) gives:

$$0 = -r_s x + D(c_s^f - c_s) \tag{4.51}$$

which upon combination with equation (4.50) and the definition of the yield coefficient directly gives:

$$x = Y_{sx}(c_s^f - c_s) \tag{4.52}$$

Thus the yield coefficient can be determined from measurement of the biomass and the substrate concentrations in the bioreactor (it is assumed that the substrate concentration in the feed flow is known).

Besides the advantage for obtaining steady-state measurements the chemostat is well suited to study dynamic conditions, because it is possible to perform well controlled transients. Thus it is possible to study the cellular response to a sudden increase in the substrate concentration by adding a pulse of the limiting substrate to the reactor or to a sudden change in the dilution rate. These experiments both start and end with a steady state and so the initial and end conditions are well characterized, and this facilitates the interpretation of the cellular response. One type of transient experiment is especially suited to determining an important kinetic parameter – namely the maximum specific growth rate. By increasing the dilution rate to a value above μ_{max} the cells will wash out from the bioreactor and the substrate concentration will increase (and eventually reach the same value as in the feed). After adaptation of the cells to the new conditions, they will attain their maximum specific growth rate and the dynamic mass balance for the biomass becomes:

$$\frac{dx}{dt} = (\mu_{max} - D)x \tag{4.53}$$

or

$$\frac{x(t - t_0)}{x(t_0)} = \exp[\,(\mu_{max} - D)(t - t_0)\,] \qquad (4.54)$$

where t_0 is the time at which the cells have become adapted to the new conditions and grow at their maximum specific growth rate. Thus, the maximum specific growth rate is easily determined from a plot of the biomass concentration versus time on a semi-log plot.

4.3.4 *The fed-batch reactor*

This operation is probably the most common in industrial processes, because it allows for control of the environmental conditions, e.g. maintaining the glucose concentration at a certain level, as well as enabling formation of very high titres (up to several hundred grams per litre of some metabolites), which is important for subsequent downstream processing. For a fed-batch reactor the mass balances for biomass and substrate are given by equations (4.38) and (4.40). Normally the feed concentration c_s^f is very high, i.e. the feed is a very concentrated solution, and the feed flow is low, giving a low dilution rate.

For the fed-batch reactor the dilution rate is given by:

$$D = \frac{1}{V}\frac{dV}{dt} \qquad (4.55)$$

and to keep D constant there needs to be an exponentially increasing feed flow to the bioreactor. Normally this is not practically possible as it may lead to limitations in oxygen supply, and the feed flow is therefore typically increased until limitations in the oxygen supply set in, after which the feed flow is kept constant. This will give a decreasing specific growth rate. However, because the biomass concentration usually increases, the volumetric uptake rate of substrates (including oxygen) may be kept approximately constant. From the above it is quite clear that there may be many different feeding strategies in a fed-batch process, and optimization of the operation is a complex problem that is difficult to solve empirically, and even when a very good process model is available calculation of the optimal feeding strategy is a complex optimization problem. In an empirical search for the optimal feeding policy the two most obvious criteria are:

- Keep the concentration of the limiting substrate constant
- Keep the volumetric growth rate of the biomass (or uptake of a given substrate) constant

A constant concentration of the limiting substrate is often applied if the substrate inhibits product formation, and the chosen concentration is therefore dependent on the degree of inhibition and the desire to maintain a certain growth of the cells. A constant volumetric growth rate (or uptake of a given substrate) is applied if there are limitations in the supply of oxygen or in heat removal.

Fed-batch cultures were used in the production of baker's yeast as early as 1915. The method was introduced by Dansk Gærindustri and is therefore sometimes referred to as the Danish method. It was recognized that an excess of malt in the medium would lead to too high a growth rate resulting in an oxygen demand in excess of that which could be met by the equipment. This resulted in the development of respiratory catabolism of the yeast leading to ethanol formation at the expense of biomass production. The yeast was allowed to grow in an initially weak medium to which additional medium was added at a rate less than the maximum rate at which the organism could use it. In modern fed-batch processes for yeast production the feed of molasses is under strict control, based on the automatic measurement of traces of ethanol in the exhaust gas of the bioreactor. Although such systems may result in low growth rates, the biomass yield is generally close to the maximum obtainable, and this is especially important in the production of baker's yeast where there is much focus on the yield. Apart from the production of baker's yeast the fed-batch process is today used for the production of secondary metabolites (where penicillin is the most prominent group of compounds), industrial enzymes, and many other products derived from cultivation processes.

4.4 Kinetic models

Kinetic modelling expresses the verbally or mathematically expressed correlation between rates and reactant/product concentrations which when inserted into the mass balances derived in section 4.3 permits a prediction of the degree of conversion of substrates and the yield of individual products at other operating conditions. If the rate expressions are correctly set up it may be possible to express the course of an entire fermentation experiment based on initial values for the components of the state vector, e.g. concentration of substrates. This leads to simulations which may finally result in an optimal design of the equipment or an optimal mode of operation for a given system. The basis of kinetic modelling is to express functional relationships between the forward reaction rates v of the reactions considered in the model and the concentrations of the substrates, metabolic products, biomass constituents, intracellular metabolites and/or the biomass concentration:

$$v_i = f_i(\mathbf{C}_s, \mathbf{C}_p, X_{macro}, \mathbf{X}_{met}, x) \tag{4.56}$$

If during the cultivation the biomass composition remains constant then the rates of the internal reactions must necessarily be proportional. This is referred to as **balanced growth** and in this case the growth process can be described in terms of a single variable which defines the state of the biomass. This variable is quite naturally chosen as the biomass concentration x (g DW l^{-1}). This is the basis of the so-called **unstructured models** which have proved adequate during 50 years of practical application to design cultivation processes (especially steady state or batch cultivations), to install suitable control devices and to estimate which process conditions are likely to give the best return on the investment in process equipment. However, these unstructured models generally have poor predictive strength and they are of little use in fundamental studies of cellular function. These models are generally referred to as structured models. The structuring of the biomass depends on what the model is to be applied for, and before we describe different models we therefore discuss aspects related to the degree of model complexity.

4.4.1   *The degree of model complexity*

A typical discussion on the mathematical modelling of biochemical systems concerns the degree of complexity – or, one could say, whether a mechanistic model or an empirical model should be applied. Especially for biological systems where we will never reach the information level where real physical-chemical models can be constructed and applied even for single processes in the cell, e.g. gene transcription, and definitely not for the overall process of cell growth. To illustrate this consider the fractional saturation y of a protein at a ligand concentration c_1. This may be described by the Hill equation (Hill, 1910):

$$y = \frac{c_1^h}{c_1^h + K} \tag{4.57}$$

where h and K are empirical parameters. Alternatively the fractional saturation may be described by the equation of Monod *et al.* (1965):

$$y = \frac{\left(La\left(1 + \dfrac{ac_1}{K_R}\right)^3 + \left(1 + \dfrac{c_1}{K_R}\right)^3\right)\dfrac{c_1}{K_R}}{L\left(1 + \dfrac{ac_1}{K_R}\right)^4 + \left(1 + \dfrac{c_1}{K_R}\right)^4} \tag{4.58}$$

where L, a, and K_R are parameters. Both equations address the same experimental problem, but whereas equation (4.57) is completely empirical with h and K as fitted parameters, equation

(4.58) is derived from a hypothesis for the mechanism and the parameters therefore have a direct physical interpretation. If the aim of the modelling is to understand the underlying mechanism of the process, equation (4.57) can obviously not be applied because the kinetic parameters are completely empirical and give no (or little) information about the ligand binding to the protein. In this case equation (4.58) should be applied, because by estimating the kinetic parameter the investigator is supplied with valuable information about the system and the parameters can be directly interpreted.

If, on the other hand, the aim of the modelling is to simulate the ligand binding to the protein, equation (4.57) may be as good as (4.58) – equation (4.58) may even be preferable because it is simpler in structure and has fewer parameters and it actually often gives a better fit to experimental data than equation (4.57). Thus, the answer to which model is preferred depends on the aim of the modelling exercise. The same can be said about the unstructured growth models (section 4.4.2), which are completely empirical, but which are valuable for extracting key kinetic parameters for growth. Furthermore, they are well suited to simple design problems and for teaching.

If the aim is to simulate dynamic growth conditions one may turn to simple structured models (section 4.4.3), e.g. the compartment models, which are also useful for illustration of structured modelling in the classroom. However, if the aim is to analyse a given system in further detail it is necessary to include far more structure in the model, and in this case one often describes only individual processes within the cell, e.g. a certain pathway or gene transcription from a certain promoter. Similarly, if the aim is to investigate the interaction between different cellular processes, e.g. the influence of plasmid copy number on chromosomal DNA replication, a single cell model (section 4.4.4) has to be applied.

Finally, if the aim is to look into population distributions, which in some cases may have an influence on growth or production kinetics, either a segregated or a morphologically structured model has to be applied (section 4.5).

4.4.2 Unstructured models

Even when there are many substrates one of these substrates is usually limiting, i.e. the rate of biomass production depends exclusively on the concentration of this substrate. At low concentrations c_s of this substrate μ is proportional to c_s, but for increasing values of c_s an upper value μ_{max} for the specific growth rate is gradually reached. This is the verbal formulation of the Monod (1942) model:

Table 4.5   Compilation of K_s values for sugars

Species	Substrate	K_s (mg l^{-1})
Aerobacter aerogenes	Glucose	8
Escherichia coli	Glucose	4
Klebsiella aerogenes	Glucose	9
	Glycerol	9
Klebsiella oxytoca	Glucose	10
	Arabinose	50
	Fructose	10
Lactococcus cremoris	Glucose	2
	Lactose	10
	Fructose	3

Values are taken from Nielsen and Villadsen (1994).

$$\mu = \mu_{max}\frac{c_s}{c_s + K_s} \tag{4.59}$$

which has been shown to correlate fermentation data for many different microorganisms. In the Monod model K_s is that value of the limiting substrate concentration at which the specific growth rate is half its maximum value. Roughly speaking, it divides the μ versus c_s plot into a low substrate concentration range where the specific growth rate is strongly (almost linearly) dependent on c_s, and a high substrate concentration range where μ is independent of c_s.

When glucose is the limiting substrate the value of K_s is normally in the micromolar range (corresponding to the mg l^{-1} range), and it is therefore experimentally difficult to determine. Some of the K_s values reported in the literature are compiled in Table 4.5. It should be stressed that the K_s value in the Monod model does not represent the saturation constant for substrate uptake, but an overall saturation constant for the entire growth process.

Some of the most characteristic features of microbial growth are represented quite well by the Monod model:

- The constant specific growth rate at high substrate concentration
- The first-order dependence of the specific growth rate on substrate concentration at low substrate concentrations

In fact one may argue that the two features which make the Monod model work so well in fitting experimental data are deeply rooted in any naturally occurring conversion process: the size of the machinery which converts substrate must have an upper value, and all chemical reactions will end up as first-order processes when the reactant concentration tends to zero. The satisfactory fit of the Monod model to many experimental data should never be misconstrued to mean that equation (4.59) is a mechanism of fermentation processes. The Langmuir rate expression of heterogeneous catalysis and the Michaelis–Menten rate expression in enzymatic catalysis are formally identical to equation (4.59), but the denominator constant has a direct physical interpretation in both cases (the equilibrium constant for dissociation of a catalytic site-reactant complex) whereas K_s in equation (4.59) is no more than an empirical parameter used to fit the average substrate influence on all cellular reactions which are pooled into the single reaction by which substrate is converted to biomass.

In the Monod model it is assumed that the yield of biomass from the limiting substrate is constant, i.e. there is proportionality between the specific growth rate and the specific substrate uptake rate. The Monod model is, however, normally used together with a maintenance consumption of substrate, i.e. the specific substrate uptake is described by the linear relation (4.18). The Monod model including maintenance is probably the most widely accepted model for microbial growth, and it is well suited for analysis of steady-state data from a chemostat (see Example 4c). Often the model is combined with the Luedeking and Piret (1959) model for metabolite production in which the specific rate of product formation is given by equation (4.17). The Luedeking and Piret model was derived on the basis of an analysis of lactic acid fermentation, and is in principle only valid for metabolic products which are formed as a direct consequence of the growth process, i.e. metabolic products formed from the primary metabolism. However, the model may in some cases be applied to other products, e.g. secondary metabolites, but this should not be done mechanically.

Example 4c. The Monod model

Despite its simplicity the Monod model is very useful for extracting key growth parameters, and it generally fits simple batch fermentations with one exponential growth phase and steady-state chemostat cultures (but rarely with the same parameters). We first consider a batch process, where substrate consumption due to maintenance can usually be neglected. In this case there is an analytical solution to the mass balances for the concentrations of substrate and the limiting substrate (Nielsen and Villadsen, 1994):

$$c_s = c_{s,0} - Y_{xs}(x - x_0) \tag{c1}$$

$$\mu_{max}t = \left(1 + \frac{K_s}{c_{s,0} + Y_{xs}x_0}\right)\ln\left(\frac{x}{x_0}\right)$$
$$- \frac{K_s}{c_{s,0} + Y_{xs}x_0}\ln\left(1 + \frac{Y_{sx}(x_0 - x)}{c_{s,0}}\right) \tag{c2}$$

Using this analytical solution it is in principle possible to estimate the two kinetic parameters in the Monod model, but since K_s generally is very low it is in practice not possible to estimate this parameter from a batch cultivation (see Figure 4.3).

For a steady state continuous culture the mass balance for the biomass, together with the Monod model, gives:

$$D = \mu_{max} \frac{c_s}{c_s + K_s} \qquad (c3)$$

or

$$c_s = \frac{DK_s}{\mu_{max} - D} \qquad (c4)$$

Thus the concentration of the limiting substrate increases with the dilution rate. When substrate concentration becomes equal to the substrate concentration in the feed the dilution rate attains its maximum value, which is often called the critical dilution rate:

$$D_{crit} = \mu_{max} \frac{c_s^f}{c_s^f + K_s} \qquad (c5)$$

When the dilution rate becomes equal to or larger than this value the biomass is washed out of the bioreactor. Equation (b4) clearly shows that the steady state chemostat is well suited to studying the influence of the substrate concentration on the cellular function, e.g. product formation, because by changing the dilution rate it is possible to change the substrate concentration as the only variable. Furthermore, it is possible to study the influence of different limiting substrates on the cellular physiology, e.g. glucose and ammonia.

Besides quantification of the Monod parameters, the chemostat is well suited to determine the maintenance coefficient. Because the dilution rate equals the specific growth rate, the yield coefficient is given by:

$$Y_{sx} = \frac{D}{Y_{xs}^{true} D + m_s} \qquad (c6)$$

or if we use equation (4.52):

$$x = \frac{D}{Y_{xs}^{true} D + m_s}(c_s^f - c_s) \approx \frac{D}{Y_{xs}^{true} D + m_s} c_s^f \qquad (c7)$$

because $c_s^f \gg c_s$ except for dilution rates close to the critical dilution rate. Equation (c7) shows that the biomass concentration decreases at low specific growth rates, where the substrate consumption for maintenance is significant compared with that for growth. At high specific growth rates (high dilution rates) maintenance is negligible and the yield coefficient becomes equal

Figure 4.4
Growth of *Aerobacter aerogenes* in a chemostat with glycerol as the limiting substrate. The biomass concentration (■) decreases for increasing dilution rate due to the maintenance metabolism, and when the dilution rate approaches the critical value the biomass concentration decreases rapidly. The glycerol concentration (▲) increases slowly at low dilution rates, but when the dilution rate approaches the critical value it increases rapidly. The lines are model simulations using the Monod model with maintenance, and with the parameter values: $c_s^f = 10$ g l^{-1}; $\mu_{max} = 1.0$ h^{-1}; $K_s = 0.01$ g l^{-1}; $m_s = 0.08$ g (g DW h)$^{-1}$; $Y_{xs}^{true} = 1.70$ g (g DW)$^{-1}$.

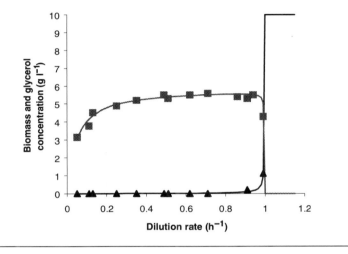

to the true yield coefficient (Figure 4.4). By rearrangement of equation (c7) a linear relationship between the specific substrate uptake rate (given by $D(c_s^f - c_s)/x$ and the dilution rate is found, and using this the true yield coefficient and the maintenance coefficient can easily be estimated using linear regression.

It is unlikely that the Monod model can be used to fit all kinds of fermentation data. Many authors have tried to improve on the Monod model, but generally these empirical models are of little value. However, in some cases growth is inhibited either by high concentrations of the limiting substrate or by the presence of a metabolic product, and in order to account for this the Monod model is often extended with additional terms. Thus for inhibition by high concentrations of the limiting substrate:

$$\mu = \mu_{max} \frac{c_s}{c_s^2/K_i + c_s + K_s} \tag{4.60}$$

and for inhibition by a metabolic product

$$\mu = \mu_{max} \frac{c_s}{c_s + K_s} \frac{1}{1 + c_p/K_i} \qquad (4.61)$$

Equations (4.60) and (4.61) may be a useful way of including product or substrate inhibition in a simple model, and often these expressions are also applied in connection with structured models. Extension of the Monod model with additional terms or factors should, however, be carried out with some restraint because the result may be a model with a large number of parameters but of little value outside the range in which the experiments were made.

4.4.3   Compartment models

Simple structured models are in one sense improvements to the unstructured models, because some basic mechanisms of the cellular behaviour are at least qualitatively incorporated. Thus, the structured models may have some predictive strength, i.e. they may describe the growth process at different operating conditions with the same set of parameters. But one should bear in mind that 'true' mechanisms of the metabolic processes are of course not considered in simple structured models even if the number of parameters is quite large.

In structured models all the biomass components are lumped into a few key variables, i.e. the vectors X_{macro} and X_{met}, which are hopefully representative of the state of the cell. The microbial activity thus becomes not only a function of the biotic variables, which may change with very small time constants, but also of the cellular composition, and consequently of the 'history' of the cells, i.e. the environmental conditions which the cells have experienced in the past.

The biomass can be structured in many different ways: in simple structured models only a few cellular components are considered whereas in highly structured models up to 20 intracellular components are considered (Nielsen and Villadsen, 1992). As discussed in section 4.4.1, the choice of structuring depends on the aim of the modelling exercise, but often one starts with a simple structured model onto which more and more structure is added as new experiments are added to the database. Even in highly structured models many of the cellular components included in the model represent 'pools' of different enzymes, metabolites or other cellular components. The cellular reactions considered in structured models are therefore empirical in nature because they do not represent the conversion between 'true' components. Consequently it is permissible to write the kinetics for the individual reactions in terms of reasonable empirical expressions, with a form which is judged to fit the experimental data with a small number of parameters. Thus Monod-type expressions are often used because they summarize some fundamental features of most cellular reactions, i.e. being first order at low substrate concentration and zero order at high substrate concentration. Despite their empirical nature structured models are normally based on some well-known cell mechanisms, and they therefore have the ability to simulate certain features of experiments quite well.

The first structured models appeared in the late 1960s from the group of Fredrickson and Tsuchiya at the University of Minnesota (Ramkrishna *et al.*, 1966, 1967; Williams, 1967), who also were the first to formulate microbial models within a general mathematical framework similar to that used to describe reaction networks in classical catalytic processes (Tsuchiya *et al.*, 1966; Fredrickson *et al.*, 1967; Fredrickson, 1976). Since this pioneering work many other simple structured models have been presented (Harder and Roels, 1982; Nielsen and Villadsen, 1992).

In these simple structured models the biomass is divided into a few compartments. These compartments must be chosen with care, and cell components with similar function should be placed in the same compartment, e.g. all membrane material and otherwise rather inactive components in one compartment, and all active material in another compartment. With the central role of the protein synthesizing system (PSS) in cellular metabolism this is often a key component in compartment models. Besides a few enzymes, the PSS consists of ribosomes, which contain approximately 60% ribosomal RNA and 40% ribosomal protein. Because rRNA makes up more than 80% of the total stable RNA in the cell, the level of the ribosomes is easily identified through measurements of the RNA concentration in the biomass. As seen in Figure 4.5, the RNA content of many different

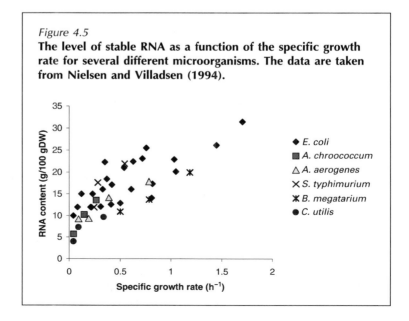

Figure 4.5
The level of stable RNA as a function of the specific growth rate for several different microorganisms. The data are taken from Nielsen and Villadsen (1994).

microorganisms increases linearly with the specific growth rate. Thus the level of the PSS is well correlated with the specific growth rate. It is therefore a good representative of the activity of the cell, and this is the basis of most of the simple structured models (see Example 4d).

Example 4d. A two compartment model
Nielsen *et al.* (1991a,b) presented a two compartment model for the lactic acid bacterium *Lactococcus cremoris*. The model is a progeny of the model of Williams with a similar definition of the two compartments:

- Active (A) compartment contains the PSS and small building blocks
- Structural and genetic (G) compartment contains the rest of the cell material

The model considers both glucose and a complex nitrogen source (peptone and yeast extract), but in the following presentation we discuss the model with only one limiting substrate (glucose). The model considers two reactions for which the stoichiometry is:

$$\gamma_{11}X_A - s = 0 \tag{d1}$$

$$\gamma_{22}X_G - X_A = 0 \tag{d2}$$

In the first reaction glucose is converted into small building blocks in the A compartment and these are further converted into ribosomes. The stoichiometric coefficient γ_{11} can be considered as a yield coefficient because metabolic products (lactic acid, carbon dioxide, etc.) are not included in the stoichiometry. In the second reaction building blocks present in the A compartment are converted into macromolecular components of the G compartment. In this process some by-products may be formed and the stoichiometric coefficient γ_{22} is therefore slightly less than 1. The kinetics of the two reactions have the same form, i.e.

$$v_i = k_i \frac{c_s}{c_s + K_{s,i}} X_A \quad ; \quad i = 1,2 \tag{d3}$$

From equation (4.13) the specific growth rate for the biomass is found to be:

$$\mu = (1 \quad 1)\begin{pmatrix} \gamma_{11} & -1 \\ 0 & \gamma_{22} \end{pmatrix}\begin{pmatrix} v_i \\ v_2 \end{pmatrix} = \gamma_{11}v_1 - (1 - \gamma_{22})v_2 \tag{d4}$$

or with the kinetic expression for v_1 and v_2 inserted:

$$\mu = \left(\gamma_{11}k_1\frac{c_s}{c_s + K_{s,1}} - (1 - \gamma_{22})k_2\frac{c_s}{c_s + K_{s,2}} \right)X_A \tag{d5}$$

Thus the specific growth rate is proportional to the size of the active compartment. The substrate concentration c_s influences the specific growth rate both directly and indirectly by determining the size of the active compartment. The influence of the substrate concentration on the synthesis of the active compartment can be evaluated through the ratio r_1/r_2:

$$\frac{r_1}{r_2} = \frac{k_1}{k_2}\frac{c_s + K_{s,2}}{c_s + K_{s,1}} \tag{d6}$$

If $K_{s,1}$ is larger than $K_{s,2}$ the formation of X_A is favoured at high substrate concentration, and it is thus possible to explain the increase in the active compartment with the specific growth rate. Consequently, when the substrate concentration increases rapidly there are two effects on the specific growth rate:

- A rapid increase in the specific growth rate, which is a result of mobilization of excess capacity in the cellular synthesis machinery.
- A slow increase in the specific growth rate, which is a result of a slow build-up of the active part of the cell, i.e. additional cellular synthesis machinery has to be formed in order for the cells to grow faster.

This is illustrated in Figure 4.6 which shows the biomass concentration in two independent wash-out experiments. In both

Figure 4.6
Measurement of the biomass concentration in two transient experiments in a glucose-limited chemostat culture with *L. cremoris*. The dilution rate was shifted from an initial value of 0.10 h⁻¹ (▲) or 0.50 h⁻¹ (■) to 0.99 h⁻¹, respectively. The biomass concentration is normalized by the steady-state biomass concentration before the step change, which was made at time zero. The lines are model simulations. The data are taken from Nielsen *et al.* (1991b).

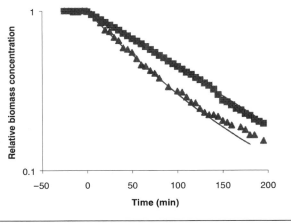

cases the dilution rate was shifted to a value (0.99 h^{-1}) above the critical dilution rate (0.55 h^{-1}), but in one experiment the dilution rate before the shift was low (0.1 h^{-1}) and in the other experiment it was high (0.5 h^{-1}). The wash-out profile is seen to be very different, with a much faster wash-out when there was a shift from a low dilution rate. When the dilution rate is shifted to 0.99 h^{-1} the glucose concentration increases rapidly to a value much higher than $K_{s,1}$ and $K_{s,2}$, and this allows growth at the maximum rate. However, when the cells come from a low dilution rate the size of the active compartment is not sufficiently large to allow rapid growth, and X_A therefore has to be built up before the maximum specific growth rate is attained. On the other hand if the cells come from a high dilution rate X_A is already high and the cells immediately attain their maximum specific growth rate. It is observed that the model is able to correctly describe the two experiments (all parameters were estimated from steady state experiments), and the model correctly incorporates information about the previous history of the cells.

The model also includes formation of lactic acid and the kinetics were described with a rate equation similarly to equation (d3). Thus, the lactic acid formation increases when the activity of the cells increases, and so it is ensured that there is a close coupling between formation of this primary metabolite and growth of the cells.

It is interesting to note that even though the model does not include a specific maintenance reaction, it can actually describe a decrease in the yield coefficient of biomass on glucose at low specific growth rates. The yield coefficient is given by:

$$Y_{sx} = \gamma_{21}\left(1 - (1 - \gamma_{22})\frac{k_2}{k_1}\frac{c_s + K_{s,1}}{c_s + K_{s,2}}\right) \tag{d7}$$

Because $K_{s,1}$ is larger than $K_{s,2}$ the last term within the parentheses decreases for increasing specific growth rates, and the yield coefficient will therefore also increase for increasing substrate concentration.

4.4.4 Single cell models

Single cell models are in principle an extension of the compartment models, but with the description of many different cellular functions. Furthermore, these models depart from the description of a population and focus on the description of single cells. This allows consideration of characteristic features of the cell, and it is therefore possible to study different aspects of cell function:

- Cell geometry can be accounted for explicitly and so it is possible to examine its potential effects on nutrient transport.

- Temporal events during the cell cycle can be included in the model, and the effect of these events on the overall cell growth can be studied.
- Spatial arrangements of intracellular events can be considered, even though this would lead to significant model complexity.

To set up single cell models it is necessary to have a detailed knowledge of the cell, and single cell models have therefore only been described for well studied cellular systems such as *Escherichia coli*, *Saccharomyces cerevisiae*, *Bacillus subtilis*, and human erythrocytes. The most comprehensive single cell model is the so-called Cornell model set up by Shuler and co-workers (Shuler *et al.*, 1979), which contains 20 intracellular components. This model predicts a number of observations made with *E. coli* and it has formed the basis for setting up several other models (Nielsen and Villadsen, 1992). Thus, Peretti and Bailey (1987) extended the model to describe plasmid replication and gene expression from a plasmid inserted into a host cell. This allowed study of host–plasmid interactions, especially the effect of copy number, promoter strength and ribosome binding site strength on the metabolic activity of the host cell and on the plasmid gene expression.

4.4.5 *Molecular mechanistic models*

Despite the level of detail the single cell models are normally based on an empirical description of different cellular events, e.g. gene transcription and translation. This is a necessity because the complexity of the model would become very high if all these individual events were to be described with detailed models that include mechanistic information. In many cases it is, however, interesting to study these events separately, and for this models where mechanistic information is included have to be used. These models are normally set up at the molecular level, and they can therefore be referred to as molecular mechanistic models. Many different models of this type can be found in the literature, but most fall in one of two categories:

- Gene transcription models
- Pathway models

Gene transcription models aim at quantifying gene transcription based on knowledge of the promoter function. The lac-promoter of *E. coli* is one of the best studied promoter systems of all, and so it is this system that has been modelled most extensively. Furthermore, this promoter (or its derivatives) is often used in connection with the production of heterologous proteins by this bacterium, because it is an inducible promoter. In a series of papers, Lee and Bailey (1984a–d) presented an elegant piece of modelling of this system, and through combination with a model for plasmid replication they could investigate, for example, the role of point mutations in the promoter on gene transcription. This promoter system is quite complex, with both activator and repressor proteins, and empirical investigation can therefore be laborious; the detailed mathematical model is a valuable tool to guide the experimental work.

In pathway models the individual enzymatic reactions of a given pathway are described with enzyme kinetic models, and it is therefore possible to simulate the metabolite pool levels and the fluxes through different branches of the pathway. These models have mainly concentrated on glycolysis in *S. cerevisiae* (Gallazo and Bailey, 1990; Rizzi *et al.*, 1997), because much information about enzyme regulation is available for this pathway. However, complete models are also available for other pathways, e.g. the penicillin biosynthetic pathway (Pissarra *et al.*, 1996). These pathway models are experimentally verified by comparing modelling simulations with measurements of intracellular metabolite pool levels – something that has only been possible with sufficient precision in the last couple of years, because it requires rapid quenching of the cellular activity and sensitive measurement techniques.

Pathway models are very useful in studies of metabolic fluxes, because they allow quantification of the control of flux by the individual enzymes in the pathway. This can be done by calculation of sensitivity coefficients (or the so-called flux control coefficients, see Box 4.4), which quantify the relative importance of the individual enzymes in the control of flux through the pathway (Stephanopoulos *et al.*, 1998).

Box 4.4 **Quantification of flux control**

In a study of flux control in a biochemical pathway the concept of metabolic control analysis is very useful (Stephanopoulos *et al.*, 1998). Here the flux control of the individual enzymatic reactions on the steady state flux J through the pathway is quantified by the so-called flux control coefficients (FCC):

$$C_i^J = \frac{v_i}{J}\frac{dJ}{dv_i}$$

where v_i is the rate of the *i*th reaction. If the enzyme concentration of the *i*th enzyme is increased its rate will normally increase, and a higher flux through the pathway may be the result. However, it is likely that due to allosteric regulation (or other regulation phenomena) there may be a very small effect of increasing the enzyme concentration. This is exactly what is quantified by the FCC, i.e. the relative increase in the steady state flux upon a relative increase in the enzyme activity. Clearly a step with a high FCC has a large control of the flux through the pathway. If a kinetic model is available for the pathway the flux control coefficients for each step can easily be calculated using model simulations. The FCCs can also be determined experimentally by changing the enzyme concentration (or activity) genetically, by titration with the individual enzymes, or by adding specific enzyme inhibitors (Stephanopoulos *et al.*, 1998).

A general criticism of the application of kinetic models for complete pathways is that despite the level of detail included they cannot possibly include all possible interactions in the system and they therefore only represent one model of the system. The robustness of the model is extremely important especially if the kinetic model is to be used for predictions, and unfortunately most biochemical models – even very detailed models – are only valid at operating conditions close to those where the parameters have been estimated, i.e. the predictive strength is limited. For analysis of complex systems it is, however, not necessary that the model gives a quantitatively correct description of all the variables, because even models that give a qualitatively correct description of the most important interactions in the system may be valuable in studies of flux control.

4.5   Population models

Normally it is assumed that the population of cells is homogeneous, i.e. all cells behave identically. Although this assumption is certainly crude if a small number of cells is considered, it gives a very good picture of certain properties of the cell population because there are billions of cells per ml medium (see Box 4.5).

Box 4.5   **Deterministic versus stochastic modelling**

In a description of cellular kinetics macroscopic balances are normally used, i.e. the rates of the cellular reactions are functions of average concentrations of the intracellular components. However, living cells are extremely small systems with only a few molecules of certain key components, and it does not really make sense to talk about 'the DNA concentration in the cell' for example, because the number of macromolecules in a cell is always small compared with Avogadro's number. Many cellular processes are therefore stochastic in nature and the deterministic description often applied is in principle not correct. However, the application of a macroscopic (or deterministic) description is convenient and it represents a typical engineering approximation for describing the kinetics in

an average cell in a population of cells. This approximation is reasonable for large populations because the standard deviation from the average 'behaviour' in a population with elements is related to the standard deviation for an individual cell through

$$\sigma_{pop} = \frac{\sigma}{\sqrt{e}}$$

Thus with a population of 10^9 cells ml^{-1}, which is a typical cell concentration during a cultivation process, it can be seen that the standard deviation for the population is very small. There are, however, systems where small populations occur, e.g. at dilution rates close to the maximum in a chemostat, and here one may have to apply a stochastic model.

Furthermore, the kinetics is often linear in the cellular properties, e.g. in the concentration of a certain enzyme, and the overall population kinetics can therefore be described as a function of the average property of the cells (Nielsen and Villadsen, 1994). There are, however, situations where cell property distributions influence the overall culture performance, and here it is necessary to consider the cellular property distribution, and this is done in the so-called **segregated models**. In the following we will discuss two approaches to segregated modelling.

4.5.1 *Morphologically structured models*

The simplest approach to model distribution in the cellular property is by the so-called morphologically structured models (Nielsen and Villadsen, 1992; Nielsen, 1993). Here the cells are divided into a finite number Q of cell states Z (or morphological forms), and conversion between the different cell states is described by a sequence of empirical metamorphosis reactions. Ideally these metamorphosis reactions can be described as a set of intracellular reactions, but the mechanisms behind most morphological conversions are largely unknown. Thus, it is not known why filamentous fungi differentiate to cells with a completely different morphology from that of their origin. It is therefore not possible to set up detailed mechanistic models describing these changes in morphology; empirical metamorphosis reactions are therefore introduced. The stoichiometry of the metamorphosis reactions is given in analogy with equation (4.4):

$$\Delta Z = 0 \tag{4.62}$$

where Δ is a stoichiometric matrix. Z_q represents both the qth morphological form and the fractional concentration (g qth morphological form (g DW)$^{-1}$). With the metamorphosis reactions one morphological form is spontaneously converted to other forms. This is of course an extreme simplification because the conversion between morphological forms is the sum of many small changes in the intracellular composition of the cell. With the stoichiometry in equation (4.62) it is assumed that the metamorphosis reactions do not involve any change in the total mass, and the sum of all stoichiometric coefficients in each reaction is therefore taken to be zero. The forward reaction rates of the metamorphosis reactions are collected in the vector **u**. Each morphological form may convert substrates to biomass components and metabolic products. These reactions may be described by an intracellularly structured model, but in order to reduce the model complexity a simple unstructured model is used for description of the growth and product formation of each cell type, e.g. the specific growth rate of the qth morphological form is described by the Monod model. The specific growth rate of the total biomass is given as a weighted sum of the specific growth rates of the different morphological forms:

$$\mu = \sum_{i=1}^{Q} \mu_i Z_i \qquad (4.63)$$

The rate of formation of each morphological form is determined both by the metamorphosis reactions and by the growth associated reactions for each form (for derivation of mass balances for the morphological forms, see Nielsen and Villadsen, 1994). The concept of morphologically structured models is well suited to describing the growth and differentiation of filamentous microorganisms (Nielsen, 1993), but it may also be used to describe other microbial systems where a cellular differentiation has an impact on the overall culture performance.

4.5.2 Population balance equations

The first example of a heterogeneous description of cellular populations was presented in 1963 by Fredrickson and Tsuchiya. In their model single cell growth kinetics was combined with a set of stochastic functions describing cell division and cell death. The model represents the first application of a completely segregated description of a cell population. In the model the cell population is described by a number density function $f(X,t)$, where $f(X,t)dX$ is the number of cells with property X being in the interval X to $X + dX$. The dynamic balance for $f(X,t)$ is given by:

$$\frac{\partial f(X,t)}{\partial t} + \frac{\partial}{\partial X}(f(X,t)v(X,t))$$

$$= 2\int_{X}^{\infty} b(X^*,t)p(X^*,X,t)f(X^*,t)dX^* - b(X,t)f(X,t) - Df(X,t) \qquad (4.64)$$

where v is the net rate of formation of the cell property X. $b(X,t)$ is the breakage function, i.e. the rate of cell division for cells with property X, and $p(X^*,X,t)$ is the partitioning function, i.e. the probability that a cell with property X is formed upon division of a cell with property X^*. Through the functions p and b a stochastic element can be introduced into the model, but these functions can also be completely descriptive. The balance of equation (4.64) was applied in the original work of Fredrickson and Tsuchiya, but in a later paper a general framework for segregated population models was presented (Fredrickson et al., 1967). Segregated models represent the complete description of a cell population and they take into account that all cells in a population are not identical. However, complete cellular segregation is rarely applied in cell culture models for two main reasons:

- For large populations, the average properties will normally represent the overall population kinetics quite well.
- The mathematical complexity of equation (4.64) is quite substantial, especially if more than a single cell property is considered, i.e. the number density function becomes multi-dimensional.

If the kinetics for product formation is not zero or first order in a given cell property, application of an average property model will, however, not give the same result as a segregated description. This is the case for production of a heterologous protein in plasmid-containing cells of *Escherichia coli*, where the product formation kinetics is not first order in the plasmid copy number. A segregated model therefore has to be applied to give a good description of the product formation kinetics (Seo and Bailey, 1985). The simplest segregated models are when the cellular property is described by a single variable, e.g. cell age, and in Example 4e the age distribution of an exponentially growing culture is derived from the general balance (4.64).

Example 4e. Age distribution model

The simplest segregated population models are those where the cellular property is taken to be described solely by the cell age a. In this case the rate of increase in the cellular property $v(a,t)$ is equal to 1. Furthermore, if it is assumed that cell division occurs only at a certain cell age $a = t_d$, the two first terms on the right-hand side of equation (4.64) become equal to zero. At steady state the balance therefore becomes:

$$\frac{d\phi(a)}{da} = -D\phi(a) \tag{e1}$$

where ϕ is a normalized distribution function:

$$\phi(a) = \frac{f(a)}{n} \tag{e2}$$

with n being the total cell number (given as the zero moment of the number density function $f(a)$). The solution to equation (e1) is:

$$\phi(a) = \phi(0)e^{-Da} \tag{e3}$$

Due to the normalization the zero'th moment of $\phi(a)$ is 1, i.e.

$$\int_0^d \phi(a)da = 1 \tag{e4}$$

which leads to:

$$\phi(a) = \frac{D}{1 - e^{-Dt_d}} e^{-Da} \tag{e5}$$

The cell balance relating to cell division (the so-called renewal equation) is given by:

$$\phi(0) = 2\phi(t_d) \tag{e6}$$

which together with equation (e3) directly gives equation (4.2). Furthermore, when equation (4.2) is inserted in equation (e5) we have the simpler expression:

$$\phi(a) = 2De^{-Da} \tag{e7}$$

Thus the fraction of cells with a given age decreases exponentially with age, and the decrease is determined by the specific growth rate of the culture (equal to the dilution rate at steady state). The average cell age is given as the first moment of $\phi(a)$:

$$\langle a \rangle = \int_0^{t_d} a\phi(a)\mathrm{d}a = \frac{1 - \ln 2}{D} \tag{e8}$$

Consequently the average age of the cells decreases for increasing specific growth rates.

References

Benthin, S., Schulze, U., Nielsen, J. and Villadsen, J. (1994) Growth energetics of *Lactococcus cremoris* FDI during energy, carbon and nitrogen limitation in steady state and transient cultures, *Chem. Eng. Sci.* **49**, 589–609.

Christensen, L.H., Henriksen, C.M., Nielsen, J., Villadsen, J. and Egel-Mitani, M. (1995) Continuous cultivation of *P. chrysogenum*. Growth on glucose and penicillin production, *J. Biotechnol.*, **42**, 95–107.

Fredrickson, A.G. (1976) Formulation of structured growth models, *Biotechnol. Bioeng.*, **18**, 1481–1486.

Fredrickson, A.G. and Tsuchiya, H.M. (1963) Continuous propagation of microorganisms, *AIChE J.*, **9**, 459–468.

Fredrickson, A.G., Ramkrishna, D. and Tsuchiya, H.M. (1967) Statistics and dynamics of procaryotic cell populations, *Math. Biosci.*, **1**, 327–374.

Galazzo, J.L. and Bailey, J.E. (1990) Fermentation pathway kinetics and metabolic flux control in suspended and immobilized *Saccharomyces cerevisiae*, *Enzym. Microb. Technol.*, **12**, 162–172.

Harder, A. and Roels, J.A. (1982) Application of simple structured models in bioengineering, *Adv. Biochem. Eng.*, **21**, 55–107.

Herbert, D. (1959) Some principles of continuous culture. *Recent Prog. Microbiol.*, **7**, 381–396.

Hill, A.V. (1910) The possible effects of the aggregation of the molecules of haemoglobin on its dissociation curves, *J. Physiol. Lond.*, **40**, 4–7.

Ingraham, J.L., Maaloe, O. and Neidhardt, F.C. (1983) *Growth of the Bacterial Cell*, Sunderland, MA: Sinauer Associates.

Lee, S.B. and Bailey, J.E. (1994a) A mathematical model for λdv plasmid replication: analysis of wild-type plasmid, *Gene*, **11**, 151–165.

Lee, S.B. and Bailey, J.E. (1994b) A mathematical model for λdv plasmid replication: analysis of copy number mutants, *Gene*, **11**, 166–177.

Lee, S.B. and Bailey, J.E. (1994c) Genetically structured models for lac promoter-operator function in the *Escherichia coli* chromosome and in multicopy plasmids: lac operator function, *Biotechnol. Bioeng.*, **26**, 1372–1382.

Lee, S.B. and Bailey, J.E. (1994d) Genetically structured models for lac promoter-operator function in the *Escherichia coli* chromosome and in multicopy plasmids: lac promoter function, *Biotechnol. Bioeng.*, **26**, 1383–1389.

Luedeking, R. and Piret, E.L. (1959) A kinetic study of the lactic acid fermentation. Batch Process at controlled pH. *J. Biochem, Microbiol. Technol. Eng.*, **1**, 393–412.

Meyenburg, K. von (1969) Katabolit-Repression und der Sprossungszyklus von Saccharomyces cerevisiae. Dissertation, ETH, Zürich.

Monod, J. (1942) *Recherches sur la Croissance des Cultures Bacteriennes*, Paris: Hermann and Cie.

Monod, J., Wyman, J. and Changeux, J.-P. (1965) On the nature of allosteric transitions: a plausible model, *J. Molec. Biol.*, **12**, 88–118.

Neidhardt, F.C., Ingraham, J.L. and Schaechter, M. (1990) *Physiology of the Bacterial Cell. A Molecular Approach*, Sunderland, MA: Sinauer Associates.

Nielsen, J. (1993) A simple morphologically structured model describing the growth of filamentous microorganisms, *Biotechnol, Bioeng.*, **41**, 715–727.

Nielsen, J. and Villadsen, J. (1992) Modelling of microbial kinetics, *Chem. Eng. Sci.*, **47**, 4225–4270.

Nielsen, J. and Villadsen, J. (1993) Bioreactors: Description and modelling, in Rehm, H.-J. and Reed, G. (eds) *Biotechnology*, Vol. 3, 2nd edn, Weinheim: VCR Verlag, pp. 77–104.

Nielsen, J. and Villadsen, J. (1994) *Bioreaction Engineering Principles*, New York: Plenum Press.

Nielsen, J., Nikolajsen, K. and Villadsen, J. (1991a) Structured modelling of a microbial system 1. A theoretical study of the lactic acid fermentation, *Biotechnol. Bioeng.*, **38**, 1–10.

Nielsen, J., Nikolajsen, K. and Villadsen, J. (1991b) Structured modelling of a microbial system 2. Verification of a structured lactic acid fermentation model, *Biotechnol. Bioeng.*, **38**, 11–23.

Olsson, L. and Nielsen, J. (1997) On-line and in situ monitoring of biomass in submerged cultivations, *TIBTECH*, **15**, 517–522.

Oura, E. (1983) Biomass from carbohydrates, in Rehm, H.-J. and Reed, G. (eds) *Biotechnology*, Vol. 3, 2nd edn, Weinheim: VCR Verlag, pp. 3–42.

Peretti, S.W. and Bailey, J.E. (1987) Simulations of host–plasmid interactions in *Escherichia coli*: copy number, promoter strength, and ribosome binding site strength effects on metabolic activity and plasmid gene expression, *Biotechnol. Bioeng.*, **29**, 316–328.

Pirt, S.J. (1965) The maintenance energy of bacteria in growing cultures, *Proc. Roy. Soc. London, Ser. B.*, **163**, 224–231.

Pissarra, P.N., Nielsen, J. and Bazin, M.J. (1996) Pathway kinetics and metabolic control analysis of a high-yielding strain of *Penicillium chrysogenum* during fed-batch cultivations, *Biotechnol. Bioeng.*, **51**, 168–176.

Ramkrishna, D., Fredrickson, A.G. and Tsuchiya, H.M. (1966) Dynamics of microbial propagation: Models considering endogenous metabolism, *J. Gen. Appl. Microbiol.*, **12**, 311–327.

Ramkrishna, D., Fredrickson, A.G. and Tsuchiya, H.M. (1967) Dynamics of microbial propagation: Models considering inhibitors and variable cell composition, *Biotechnol. Bioeng.*, **9**, 129–170.

Rizzi, M., Baltes, M., Theobald, U. and Reuss, M. (1997) In vivo analysis of metabolic dynamics in *Saccharomyces cerevisiae*: II. Mathematical model, *Biotechnol. Bioeng.*, **55**, 592–608.

Roels, J.A. (1983) *Energetics and Kinetics in Biotechnology*, Amsterdam: Elsevier Biomedical Press.

Seo, J.-H., Bailey, J.E. (1985) A segregated model for plasmid content and product synthesis in unstable binary fission recombinant organisms, *Biotechnol. Bioeng.*, **27**, 156–165.

Shuler, M.L., Leung, S.K. and Dick, C.C. (1979) A mathematical model for the growth of a single bacterial cell, *Ann. NY Acad. Sci.*, **326**, 35–55.

Sonnleitner, B. and Fiechter, A. (1988) High performance bioreactors: A new generation, *Anal. Chim. Acta*, **213**, 199–205.

Stephanopoulos, G., Nielsen, J. and Aristodou, A. (1998) *Metabolic Engineering*, San Diego: Academic Press.

Tsuchiya, H.M., Fredrickson, A.G. and Aris, R. (1966) Dynamics of microbial cell populations, *Adv. Chem. Eng.*, **6**, 125–206.

Williams, F.M. (1967) A model of cell growth dynamics, *J. Theor. Biol.*, **15**, 190–207.

5 Microbial Synthesis of Commercial Products and Strain Improvement

Iain S. Hunter

5.1 Introduction

Although the use of microorganisms in the production of commercially useful products is well established (see Chapter 1 for more details), the general public have yet to fully appreciate the microbial origin of some well-known commodities. For example, citric acid, an organic acid which chemists found very difficult – if not impossible – to synthesize with the correct stereospecificity, is now widely produced through fermentation by either *Aspergillus niger* or the yeast *Yarrowia lipolytica*. It was the expertise gained with such fermentations that allowed the rapid development of the antibiotics industry in the early 1940s (see below), with a consequential leap in quality of healthcare.

In our contemporary lifestyle, the microbial origins of many products are not well recognized. For example, xanthan gum, a product made by *Xanthomonas campestris*, is widely used in the manufacture of ice-cream as it prevents the foams from collapsing with time. In a slightly modified form, xanthan gum gives paints their unique property of sticking to the brush without dripping but spreading smoothly and evenly from the brush to the surface.

In addition to natural products such as the chemicals mentioned above, recombinant strains of *Escherichia coli*, constructed by genetic engineering, are now used for the production of 'nonnatural products' such as human insulin. Another example is biological washing powders which contain enzymes drawn from different species, particularly those belonging to the genus *Bacillus*. Microbial products thus have a significant impact on our everyday lives.

Lastly, the 'new biotechnology' industry is critically dependent on microbial fermentations for the production of the extremely high-value recombinant therapeutic products, e.g. human insulin and human growth hormone, as well as the low-value recombinant products, e.g. prochymosin which provides an ethical alternative to rennin (chymosin) in the manufacture of cheese.

For each useful microbial product, it is necessary to develop a strain which, in addition to demonstrating a high rate of production, fulfils a number of important criteria including the ability to utilize effectively the cheapest available substrate, the ability to complete fermentation in a relatively short time and to produce a final product that is relatively pure so that downstream processing costs are kept to a minimum. Thus, strain improvement programmes must be undertaken to achieve these targets. This chapter has two main themes: (1) the use of classical strain improvement strategies, which have been extremely successful in the past, and continue to provide a foundation to develop strains which have the potential to make elevated levels of new products, especially in the antibiotics industry and (2) the use of recombinant (cloning) techniques in the improvement of existing products and development of new therapeutic products of extremely high value.

5.2   The economics and scale of microbial product fermentations

The type of fermentation used, as well as its size, duration and nutrient profile, will depend critically on the nature of the microbial product. For 'low-value, high-volume' products, such as citric acid and xanthan gum, high capacity fermentors (often up to 800 m^3 in volume) will be used. The duration of the fermentation and costs of the nutrients and 'utilities' (heating, cooling and air) are the critical factors in the overall profitability of this business.

'Medium-value, medium-volume' products, such as antibiotics, are typically made in fermentors that are considerably smaller (100–200 m^3), but for which the duration and utility/nutrient costs will still be a significant factor.

'High-value, low-volume' products, such as the recombinant therapeutic proteins, are made in small (approximately 400 l, i.e. 0.4 m^3) fermentors for which the cost of the nutrients and utilities are a minor factor in the overall feasibility and profitability.

For all but the highest-value products, nutrient costs (especially of the main carbon source) are critical. Depending on the vagaries of world commodity markets, complicated further by artificially imposed trade tariffs when the nutrients are shipped across frontiers, the costs and availability of nutrients can fluctuate at alarming rates. Flexibility in the choice of nutrient used for any fermentation process is of paramount importance, and strain improvement programmes must accommodate this important factor.

LOW-VALUE, HIGH-VOLUME PRODUCT
A commercial term for a product whose cost is low (of the order of £6/kg, or less) but which is required in vast quantities (many millions of kg per year). Citric acid and xanthan fall into this category.

MEDIUM-VALUE, MEDIUM-VOLUME PRODUCT
A commercial term for a product which falls between the extremes of a low-value, high-volume and high-value, low-volume product. Typically, antibiotics fall into this category, being required on a scale of up to 100 000 kg per year at a cost of around £60 per kg.

HIGH-VALUE, LOW-VOLUME PRODUCT
A commercial term for a product (usually biologically active) which is extremely potent and is required only in small (kg) quantities throughout the world per year. However, the price charged for the product may be high (up to £60/mg). Recombinant proteins, such a human insulin, interferon and growth hormone fall into this category.

By contrast, for the high-value (recombinant) products, the emphasis of the strain improvement programme is on the stability of the strain, expression levels, and overall quality of product, rather than the cost of the fermentation process *per se*.

A fermentation process constitutes a business which aims to sell the product at an overall profit. This issue positions strain improvement programmes at the interface between science and the commercial world, and requires a different set of criteria to judge whether a task is worthwhile or not. Proposals for strain improvement programmes must first be vetted to establish the costs of the research and development that will be necessary to achieve the stated goal, and to set these costs against the annual savings likely to ensue, should that piece of work deliver the expected gains in productivity. It often comes as a shock to researchers new to this field that projects which are highly innovative, but to which some risk of failure is attached, are not funded because the return on such an investment (set against the risk) is not high enough. For example, a strain improvement programme for a 'mature product' would be difficult to justify if the annual R&D cost could not be recouped in around three years through the projected increase in productivity.

In other words, the annual cost of strain improvement programmes must be only around three times (or less) the projected annual cost savings. Acclimatization to such rigorous reviews of research plans and draconian decision-taking constitutes a sharp learning curve for newly-recruited research staff.

5.3 Different products need different fermentation processes

At the 'low-value' end of the microbial products business, the margins on profitability are extremely tight. For citric acid, the titre must be greater than 100 g l^{-1}, with carbon conversion efficiency (amount of substrate converted to product on a 'per gram' basis) close to 100% for the process to become economic. Citric acid is produced concurrently with microbial growth, but a fine balance has to be struck between the amount of cells that are made in the fermentation (which consumes some of the carbon source that otherwise could be used to make product) and the fact that a doubling of the cell mass may result in twice the volumetric rate of citric acid production (as each cell acts as its own 'cell factory'), but may hinder the eventual purification of the product. Growth can be arrested by, typically, limiting the amount of nitrogen available to the culture, in which case citric acid is still produced for some time by the cells in stationary phase. Cheap carbon sources (e.g. unrefined molasses), fast production runs with minimal 'turn around' time, and a cheap and

MATURE PRODUCT
A commercial term for a product which has been in the market-place for some time and for which the cost of manufacture is very important. By contrast, a new product is either patented (in which case the company that owns the patent has sole rights to market it for up to 17 years) or has some novel property compared with its competitors. In either case the product can be sold at a premium price. However, as other competitive products appear on the market or, in the case of a patented product, when the patent runs out, the selling price comes under pressure and the profit associated with it is reduced. With such a 'mature product', the cost of manufacture (amount of product made and efficiency of the overall process) becomes critical.

rapid means of extracting the product are most important, and the strain improvement programme of this 'mature' product will be focused on substrate flexibility and rapid production.

Medium-value products, such as antibiotics, are produced by microbes which, when first isolated from the soil, usually make detectable, but vanishingly small, quantities (a few micrograms) of the bioactive substance. Most commercial fermentation processes will only become economically viable if the strain is eventually capable of making more than $15–20$ g l^{-1} of the bioactive product using comparatively cheap carbon sources, such as rape seed oil, and low grade sources of protein, such as soya bean or fish meal. Antibiotic production (see below) usually occurs following the onset of the stationary phase, i.e. after growth has been arrested, but many antibiotics contain nitrogen as well as carbon, so growth is limited by the supply of phosphate. A steady supply of nutrients containing carbon and nitrogen, but not phosphate, is fed to the fermentation during the phase of antibiotic production.

Once a new microbial metabolite has been discovered, a campaign of 'empirical strain improvement' is undertaken to boost the level of production to a titre at which the process becomes economically viable.

For 'high-value' products, such as recombinant therapeutic proteins, the cost of the fermentation broth is not a major issue. To ensure consistency of the final product, expensive well-defined media (either Analar mineral medium or high-specification tryptone or hydrolysed protein) are used along with a high purity carbon source (usually glucose). Following the onset of the stationary phase, the production of high-value products is triggered, generally in response to an external signal such as temperature shift or the addition of an exogenous gratuitous inducer. The levels of production may be relatively modest (less than 1 g l^{-1}) to meet commercial targets, but higher levels are always sought. Because the products are often formulated into injectable medicines, in which there is a fear of raising an immune response if the recombinant protein is not 100% pure and folded into the correct three-dimensional structure, a major emphasis of strain improvement programmes addresses this important regulatory issue, rather than the cost of the media and utilities involved in the process.

5.4   Fed-batch culture as the paradigm for many efficient microbial processes

Simple batch culture is a fairly inefficient way to synthesize a microbial product, as a substantial proportion of the nutrients present in the fermentation are used to make the biomass and the opportunity then to use the biomass as a cell factory to make the product is limited to the time of the growth period. Although very high levels of productivity can often be achieved in continuous culture, this technique has the disadvantages that large volumes of medium (often expensive medium) are required, and the product is made in a dilute stream which has to be concentrated before final isolation can take place. Fed-batch culture is a 'half-way' approach. The cells are grown up in batch culture and then the resident biomass, which is no longer growing, is dedicated to product formation by feeding nutrients, except for that which is chosen to limit growth. This type of culture technique was developed by the antibiotics industry.

To understand the nature of antibiotic fermentations, it is important to take account of the life cycle of the producing microorganisms within the natural ecosystem. This example is from *Streptomyces*, the filamentous bacteria that make the majority (> 60%) of natural antibiotics, including streptomycin, the tetracyclines and erythromycin. The life cycles of the eukaryotic fungi, which produce the natural penicillins, are broadly similar. The cycle begins with the spore, which may lie dormant in the soil for many years. When nutrients and water become available, the spore germinates and outgrowth takes place (Figure 5.1). These organisms are said to be 'mycelial' in nature, as they colonize soil particles (or the agar in petri dishes, if they are in the laboratory) by extending outwards in all directions (radial growth) in a fixed branched pattern. This is often called **vegetative mycelium**. Inevitably, at some point, nutrients or water become in short supply and it is at this point that they differentiate, ultimately to form spores again. Initially, the differentiation process involves the formation of **aerial mycelium** in which the biomass no longer extends out radially, but rises up away from the plane of the radial growth and forms elaborate coiled structures which then septate after a while and the spores are formed again. Antibiotics are made at the time of differentiation between the vegetative mycelial and the aerial mycelial stages. This may seem a paradox, as the signal to undertake this differentiation step is lack of nutrient or water – how can a microorganism starved of nutrients elaborate such a complex structure as the aerial mycelium? The answer is that extracellular lytic enzymes are made and released at the same time as the **differentiation switch**.

These enzymes digest the vegetative mycelium, in part, and the cellular building blocks that are released are used to construct the aerial mycelium. In this way, part of the vegetative mycelium is 'sacrificed' to allow the aerial mycelium to be formed and sporulation to proceed. The process can be viewed as a survival strategy because, when the spores have formed, the

DIFFERENTIATION SWITCH

A genetic switch that changes the pattern of the genes that are expressed, in such a way that the morphology of the cell changes (differentiates) along with some of the functions that it performs. The step cannot be reversed easily, i.e. the cells do not return to their original shape and functionality in a single step.

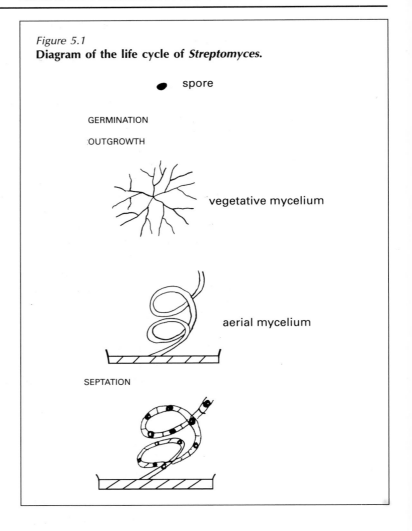

Figure 5.1
Diagram of the life cycle of *Streptomyces.*

organism's DNA is held in an inert state so that the life cycle can begin again when water and nutrients are plentiful. It has been suggested that the reason that these microorganisms make antibiotics exactly at the same time as this critical differentiation step is to sterilize the micro-environment around the lysing vegetative mycelia, so that other microbial predators cannot have an advantageous feed on the cellular nutrients released (a view that is not held by all scientists in the field).

When Man wishes to harness the vast natural potential of the *Streptomycetes* or the eukaryotic fungi to make antibiotics in quantities that would be astounding in the natural ecosystem, the organisms have to be grown in liquid culture and large volumes. Care has to be taken that, in transferring the organisms from growth and antibiotic production on solid substrata to liquid media, the essential features of the biology of the anti-biotic production process are not lost. In liquid medium, it is very unusual for aerial mycelia to be made and sporulation is observed even less frequently. Despite this, it is possible to induce the cultures to make an antibiotic in liquid culture.

The physiology of the fermentation process has to be adapted to suit the biology of the microbe. As one of the triggers for differentiation in the natural ecosystem is deprivation of nutri-ents, then the same strategy may also be applied to the fermen-tation process. All antibiotics are composed of carbon atoms and many contain nitrogen atoms. Therefore it would be a poor tactic to attempt to trigger the onset of antibiotic production by limiting the cells' supply of either of these elements, because after antibiotic production had commenced, the cells would be starved of one of the most important chemical elements needed to biosynthesize the antibiotic structures. Fortunately, very few antibiotics contain phosphorus in their elemental composition, so the cells are most often limited by the supply of phosphate to trigger antibiotic production. If the other nutrients are supplied in sufficient quantities (but not in vast excess, see later) then antibiotic production will continue for many days (often 8–10 days) at a rate which is linear with time (Figure 5.2). Eventually, this rate starts to tail off and it makes economic sense to termi-nate the fermentation at this point, recover the maximal amount of product in the shortest possible time, and then prepare the vessel for another round of fermentation.

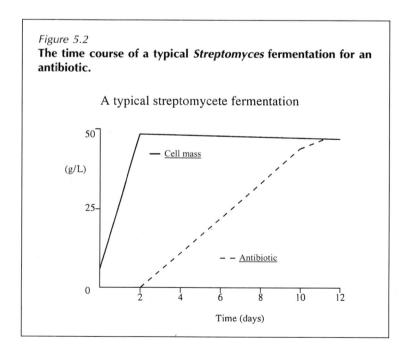

Figure 5.2
The time course of a typical *Streptomyces* fermentation for an antibiotic.

Fermentations for high-value recombinant therapeutic proteins are best undertaken by fed-batch cultures, which maximize
the use of the biomass factory and deliver the product in the
most concentrated form. However, the duration of the production phase for these high-value fermentations is considerably
shorter than for the antibiotic fermentations – typically around
8 hours, or less.

5.5   Tactical issues for strain improvement programmes

In practice, the order in which the experimental strategies for
improving strains is used depends on the nature of the product
and of the fermentation. (In the initial stages of developing a
strain from a 'soil isolate' to become a commercial producer of
an antibiotic, random mutagenesis of a population of the producer microorganism is undertaken (see below) and the progeny
of the mutagenic treatment are screened for higher levels of the
antibiotic. Subsequently, more directed screens are used to further enhance the titre of the antibiotic.

By contrast, improvement of strains making recombinant
therapeutic proteins starts with some very directed strategies
and finishes with a more random, empirical approach. This is
because, in the initial stages of strain improvement, there is a
well-defined template of experimental improvements that can be
followed. Subsequently, the performance of the strain can be
improved further by the empirical approach.

5.6   Strain improvement: the random, empirical approach

In this approach a large number of cells in the population is
subjected to a mutagenic treatment (Figure 5.3). A population
of the microorganism is treated with a mutagen (an agent that
will induce changes in its DNA). This is often the chemical car-
cinogen, nitrosoguanidine (NTG), but other mutagenic agents
such as UV light or caffeine may be used. The treatment is tailored so that each cell, on average, has a single mutation induced. The mutagenized population is plated on agar which will
support the growth of the normal ('wild-type') strain. Some
mutations will be very deleterious to the growth of the organism
and cells containing these will not grow up to become colonies.
Those colonies (each originally derived from a single cell) that
show little or no growth impairment are then tested randomly
to estimate how much antibiotic they can make. Classically, this
was done by overlaying the agar plate containing the surviving

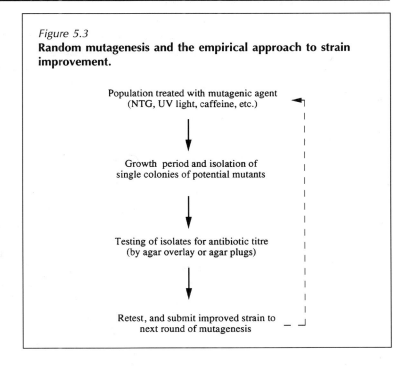

Figure 5.3

Random mutagenesis and the empirical approach to strain improvement.

Population treated with mutagenic agent
(NTG, UV light, caffeine, etc.)

↓

Growth period and isolation of
single colonies of potential mutants

↓

Testing of isolates for antibiotic titre
(by agar overlay or agar plugs)

↓

Retest, and submit improved strain to
next round of mutagenesis

colonies after mutagenesis with soft agar containing another microorganism that is sensitive to the antibiotic. Colonies producing more antibiotically-active substance will produce larger 'zones of inhibition' of growth of the culture in the overlay. However, this technique does not take into account the fact that some colonies may be larger (i.e. have a greater diameter) than others. To circumvent this problem, a cylinder of the colony and the agar underneath is cut out (usually with a cork borer) and the 'agar plug' placed on a fresh lawn of sensitive bacteria spread on a new agar plate. The cylinders are uniform and contain comparable amounts of cell material. This modification takes account of differences in sizes of different mutagenized colonies. The whole procedure is tedious but has to be undertaken in a painstaking manner. In recent times, the pharmaceutical industry has used robots to perform the tasks required in random screening. There has been a tendency to move away from screening for how much antibiotic is made by colonies on a plate, to liquid media-based cultures which reflect the situation in fermentors more accurately. It is not unusual for a screening robot to evaluate the performance of over a million mutants in a year – such 'high throughput screening' is orders of magnitude more efficient than using people for these tasks. However, the important human factor then becomes how well the screening operation can be designed and run!

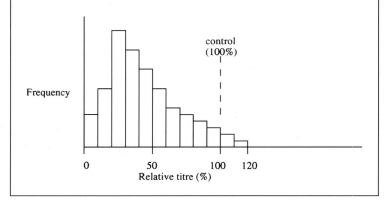

Figure 5.4
Antibiotic titres of individual survivors after random mutagenesis. The histogram shows the frequency distribution of titres of survivors after random mutagenesis, for a strain whose lineage has been developed for around 2 years. Notice that only a few isolates perform better than the control (some of this may be due to error in determining the titre) and that most survivors perform much worse than the control, as the mode is about 25%.

Most of the progeny will make lower levels (Figure 5.4) of the microbial product and only a few will have larger titres, often showing 10% (or less) improvement. The reason for this is that most mutations are deleterious to production and only a few mutations will result in slightly elevated levels of production. It is very unusual to isolate mutants that have large (> 20%) increases in titre.

These individual isolates are then tested in small-scale fermentations to confirm that they really are improved mutants. Most turn out not to have improved titres, and are discarded. Mutants that do show some promise will then be tested intensively at the laboratory and pilot plant levels.

Different companies have different strategies at this stage. Some may carry out extensive laboratory trials at around the 10 l level and scale up the mutants directly to the large production fermentors in one step. Other companies may adopt a more conservative posture and scale up the volume of fermentation in stages. For a process that is undertaken at 100 000 l production scale, the stages in scale up would typically be from 10 l to 500 l, then onward to 5000 l and finally to the large fermentors. The doctrine is that operation of these large fermentors is expensive – both in the costs of fermentation broth and in lost opportunity, as that large fermentor might otherwise have been used to make another valuable product. The view of those fermentation technologists who scale up in one step is that time is wasted solving problems associated with the intermediate stages, and that a production fermentation that is modelled well at the bench scale should, by definition, predicate immediate translation of the new mutant and its process details to the production fermentor hall.

It is often overlooked that the initial improvement in titre which is achieved with a new mutant at the production scale can often be improved still further by 'process optimization' through adjustments to the media and fermentation conditions that better suit the physiology of the new mutant. Process optimization invariably makes an equal, if not greater, contribution to overall increase in titre than the stepwise increase in fermentation performance that is achieved when the new mutant is adopted initially at the production scale.

5.7   Strain improvement: the power of recombination in 'strain construction'

In the early stages of undertaking a random mutagenesis programme for a new product (i.e. starting from a new soil isolate which makes a small amount of product), there are likely to be rapid, concurrent advances in the performances of several strains. These strains, separately, will have individual desirable properties. For example in an antibiotic programme, one strain may give a higher titre, while another may use less carbon source to make a level of antibiotic equivalent to that of the parent (and so be more economical). Yet another strain may show no better performance in terms of titre of product or the amount of nutrients consumed to make it, but may carry out the fermentation at a lower viscosity, which will make it easier for the product to be extracted from the fermentation broth and purified. Different 'lineages' of the strain are developed very rapidly in these early days in the desire to improve the fermentation considerably. However it soon becomes desirable to construct strains with a combination of several of the desirable properties already present in the individual lineages. To achieve this, desirable traits from each lineage must be 'recombined' genetically into a single strain which possesses all of the desirable properties, or 'traits'.

Most microbial species are able to exchange genetic information with other members of their species by recombination. Construction of the genetic maps that appear in many textbooks is based on exploitation of these natural recombination processes. For *E. coli* and other enteric Gram-negative bacteria, protocols to undertake genetic recombination are well documented and depend on plasmid-based mobilization of the bacterial chromosome. However, the producers of many microbial metabolites,

Figure 5.5
Strain construction by recombination. In this example two lineages have been derived from the first soil isolate. One lineage shows a steady improvement in titre with each new strain, but this is accompanied by wasteful utilization of carbon source. The other lineage makes very little antibiotic, but does so in a very efficient way. By genetic recombination of these 'traits', the attributes of both strands of the culture lineage can be combined together in a single strain.

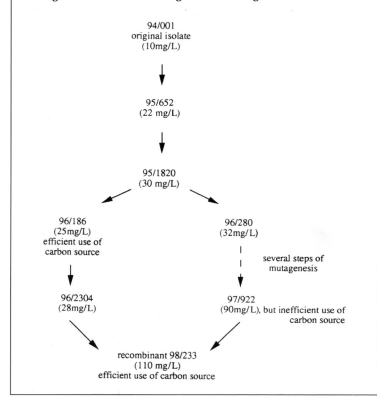

including the *Streptomyces*, have a life cycle that involves a sporulation stage (described above). With these species, the most effective method of constructing recombinants that have combinations of several desirable traits is to undertake 'spore mating': to mix spores of the parents with the individual traits, germinate them together, allow them to go through an entire life cycle (Figure 5.1), and (from the spores formed at the end of this life cycle) to select those which have exchanged *some* genetic information (i.e. select those that show that they have undertaken some degree of recombination) and then to screen them to identify the individuals that have received the desired combination of traits (Figure 5.5).

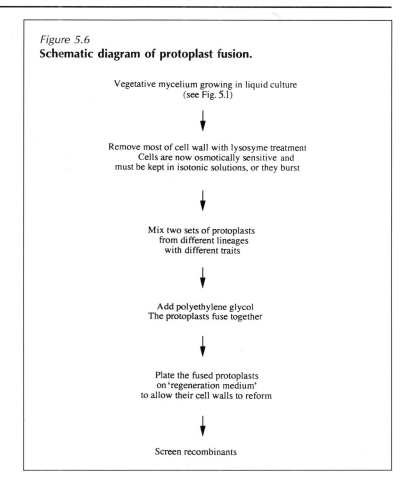

Figure 5.6
Schematic diagram of protoplast fusion.

Vegetative mycelium growing in liquid culture
(see Fig. 5.1)

Remove most of cell wall with lysosyme treatment
Cells are now osmotically sensitive and
must be kept in isotonic solutions, or they burst

Mix two sets of protoplasts
from different lineages
with different traits

Add polyethylene glycol
The protoplasts fuse together

Plate the fused protoplasts
on 'regeneration medium'
to allow their cell walls to reform

Screen recombinants

This screening process is usually conducted empirically, in the same way as screening of strains after random mutagenesis. Unfortunately, a major drawback with this procedure can often arise: as strains with further improved titres are selected and the strain lineage becomes longer, the ability to sporulate is often lost. Although sporulation is associated with the differentiation process, antibiotic production can often become decoupled from sporulation in the advanced high-titre strains. The strains become asporogenous, which makes such a 'spore mating' strategy impossible. For some strains, the problem can be circumvented by undertaking strain construction by recombination between the vegetative mycelia of different parents. However, not all strains are amenable to this approach and for them, a more involved strategy called **protoplast fusion** (Figure 5.6) has to be enacted. The most common pitfall with protoplast fusion is the time taken to develop the media conditions and culture techniques that allow the protoplasts to regenerate satisfactorily back into fully-competent mycelium (see Figure 5.6). Often an extensive series of empirical range-finding experiments has to be undertaken to optimize the conditions for regeneration, and occasionally it proves impossible to develop a protocol for a particular species.

5.8 Directed screening for mutants with altered metabolism

Although the random mutagenesis and recombination approaches are fruitful in acquiring mutants which give initial improvements in titres, in due course as the level of microbial product produced increases, it becomes more difficult (in parallel) to isolate further improved strains using the empirical, random approach. For example, in the development of an antibiotic fermentation process, by this time the titre of product will have reached a few grams per litre and considerable background data will have been gathered on the physiology and biochemistry of the fermentation process.

These data can be analysed to diagnose whether wasteful metabolites are being made during the fermentation process (see Chapters 6 and 7 for further details). For example, many *Streptomycete* fermentations often consume copious quantities of glucose and excrete pyruvate and α-*oxo*glutarate, giving low conversion yields of glucose into product. If the over-utilization of glucose can be prevented by careful process control (e.g. limiting the supply of glucose) then the economics of the process will improve substantially. However the same objective can be achieved by generating mutants which will use the glucose less quickly. Such a strategy is amenable to a 'directed screening' approach.

In the case of uncontrolled uptake and wasteful metabolism of glucose, it is known that mutants that are resistant to 6-deoxyglucose (a toxic analogue of glucose) have reduced or impaired glucose uptake and subsequent metabolism. In this strategy, a mutagenic treatment is performed on a population of cells or spores, in the same way as described in Figure 5.3. However, instead of plating all of the survivors under non-selective conditions, the survivors are plated directly onto a nutrient agar to which the desired selective pressure can be applied. In this case, the survivors would be plated on media containing 6-deoxyglucose at a level that, under normal conditions, just prevents the growth of the microorganism. The media would also contain a second carbon source (usually glycerol) on which the cultures can normally grow quite vigorously. Out of the millions of survivors of the initial mutagenic treatment, only those which are altered in some aspect of their metabolism of 6-deoxyglucose will survive and grow on such a selective medium. These isolates can then be screened conventionally to identify the individuals amongst the population which may have the potential in the fermentor to display a reduced level of glucose uptake and consequently a more balanced metabolism of the sugar, which does not involve the wasteful production of pyruvate and the α-oxo-organic acids.

Many survivors that have become resistant to 6-deoxyglucose will have mutated to confer complete exclusion of the toxic analogue from the cell, but in addition, they will no longer be able to utilize glucose at all and will have to be discarded. Strains which are totally incapable of glucose utilization are of no use to the fermentation industry which is substantially based on cheap, glucose-rich carbon sources.

However, the minor class of analogue-resistant survivors, which have impaired, but still significant, glucose uptake are the sought-after prizes in this 'directed screen'. The terminology 'directed' is appropriate because the researcher defines the exact conditions under which such mutants should survive – to generate new culture isolates with improved fermentation performance through less wasteful metabolism of glucose.

Directed screening is an extremely important tactic that may be employed to great advantage after the basic details of the fermentation have been worked out and understood. The value of directed screening is the reduction in numbers of mutants to be screened (typically by around 10 000-fold) before an isolate with the desired characteristics is identified. Such a reduction in workload allows time to establish the metabolite production profile of each survivor and to evaluate the data in greater depth. Such is the power of the directed screen that rare spontaneous mutants may be isolated, rather than those generated by mutagenic agents. If 10^6 to 10^8 cells are plated on a single plate containing a toxic analogue, a few spontaneous mutants will always be isolated.

In addition to carbon utilization, the flux of nitrogen (fixation and metabolism of ammonia) and phosphorus (phosphate metabolism) can be altered by specific mutations, for which directed screens can be devised to select for potentially improved mutants. Fixation of ammonia takes place via the enzyme glutamine synthetase in virtually all microbes. The flux through this step is often altered in mutants that are resistant to bialaphos, a toxic compound that specifically inhibits glutamine synthetase. Mutants with altered phosphate metabolism may be obtained by directed screening for resistance to toxic analogues, such as arsenite or dimethyl arsenite.

Fluoroacetate is a classic metabolic inhibitor which poisons the tricarboxylic acid (TCA) cycle by being converted to

Figure 5.7
Diagram of the biosynthetic origin of isopenicillin N. This has three component amino acids, condensed together with peptide bonds. The origins of the amino acids are shown between the dotted lines.

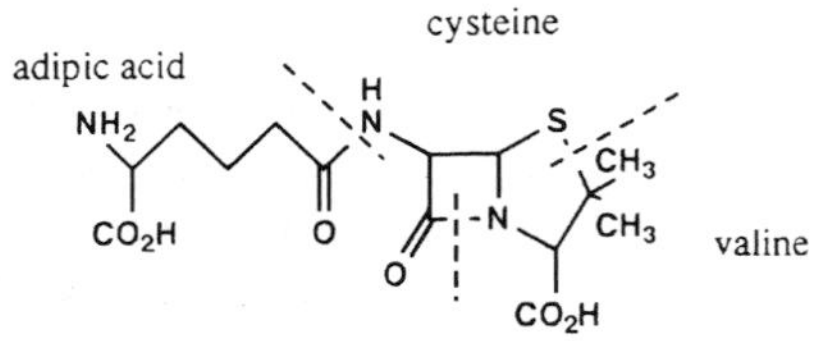

fluorocitrate, a toxic analogue of citric acid. Mutants which are more resistant (or sometimes those which are more sensitive) to fluoroacetate often have a TCA cycle with altered properties. As the TCA cycle is a fundamental component of cellular aerobic metabolism (conditions under which most fermentations are conducted), then these mutants can have important properties and be fruitful sources of improved strains.

The building blocks for all microbial products, including the antibiotics, are common metabolic precursors used in other biosynthetic processes of the cell. For example, the three components of penicillin are two amino acids, cysteine and valine, together with adipic acid, which is a precursor of lysine. Thus, penicillins can be viewed biosynthetically as a simple but modified tripeptide (Figure 5.7) of three common amino acids. During microbial growth, cellular metabolism is painstakingly controlled to ensure that supplies of all 20 amino acids that are needed for growth are available in a balanced fashion.

In a fed-batch culture, such as an antibiotic fermentation, tightly regulated metabolism during the growth phase is followed by the production phase (Figure 5.2), during which the commercial aim is to produce a single product quickly and at high levels – to the exclusion of others. As this microbial product will probably be made from a few key metabolic intermediates (e.g. during production of penicillin, only a supply of the three amino acids will be in high demand), then metabolism must be altered to satisfy this increased demand, while minimizing the side reactions of wasteful metabolism. Directed screens can be devised which decouple the usual control strategies of the biosynthetic pathways (such as feedback inhibition and cross-pathway regulation) which normally keep the supply of all precursors just balanced to the needs of the growing cell.

By way of example, consider the supply of adipic acid (part of the lysine pathway) for the biosynthesis of penicillin (Figure

5.7). Normally, the lysine pathway is subject to end-product feedback inhibition. The toxic analogue of lysine, δ-(2-amino-ethyl)-L-cysteine, also inhibits the first step of lysine biosynthesis. Mutants that are resistant to this toxic analogue are no longer subject to end-product feedback inhibition of the early part of the biosynthetic pathway. They have an enhanced flux of precursor supply to adipic acid and often produce higher titres of penicillin.

Almost every metabolic pathway that supplies the precursors of microbial products is amenable to this type of directed screening strategy. Thus the rate of 'fuel supply' for biosynthesis of microbial products can be enhanced.

The last example of directed screening relates to the selection of mutants which produce elevated levels of the enzymes responsible for catalysis of the precursor building blocks for the biosynthesis of the backbone structures of antibiotics, such as the tetracyclines and erythromycin. They are polymers of acetyl-CoA and methylmalonyl-CoA respectively, and both are made by a process which is essentially the same as that for fatty acid biosynthesis. The antibiotic cerulenin targets the enzyme complexes responsible for fatty acid biosynthesis and acts to starve the growing cell of the fatty acids necessary for insertion into the membrane. Mutants which are resistant to toxic levels of cerulenin have circumvented this problem by making elevated levels of the fatty acid biosynthetic enzymes to 'titrate out' the effect of the antibiotic. The close similarity between the enzymes of fatty acid biosynthesis and those which make the backbones of erythromycin and tetracycline allows cerulenin to inhibit the biosynthesis of these antibiotics. Mutants which can still make the antibiotic in the presence of cerulenin have elevated levels of the biosynthetic enzyme complexes. When cerulenin is removed, they retain the high level of biosynthetic capability which, if supplied with enough of the metabolic precursor 'fuel', results in higher titres of the antibiotic. The utility of the directed screening approach is that the survivors of the mutagenic treatment are invariably altered in some aspect of fatty acid or antibiotic biosynthesis.

Individual mutants made by directed screening approaches can be recombined together using genetic recombination, spore mating or protoplast fusion, as described above, to combine several desired traits.

5.9 Recombinant DNA approaches to strain improvement for low- and medium-value products

The advent of recombinant DNA techniques, first devised for *E. coli* in the 1970s, has meant that new strategies can be applied to strain improvement for low- and medium-value microbial products. It has taken some time for recombinant techniques to be developed and applied to the commercial strains, such as the filamentous bacteria and filamentous fungi which are the mainstay of this sector of the fermentation industry.

In addition, regulatory hurdles have to be crossed to gain approval to undertake fermentations at the production scale with these 'genetically-engineered microorganisms'. Gaining such approval is time-consuming and costly. Thus, there have to be good long-term economic reasons for adopting a recombinant DNA strategy for strain improvement.

In the case of antibiotic fermentations, temporal expression of antibiotic biosynthesis is regulated tightly as part of the cellular differentiation pathway (Figure 5.2). Production of antibiotics is costly to the cell in terms of carbon and energy. Therefore, it is hardly surprising that tight control of expression has evolved. The genes for antibiotic biosynthesis are invariably clustered together, irrespective of whether the microbe is a prokaryote or eukaryote. In *Streptomyces*, regulation of expression of such gene clusters is controlled by a **positive activator** – a **master gene** which, when it is switched on, makes a protein that targets itself to the various promoters of the gene cluster and switches them on in concert (Figure 5.8).

Figure 5.8
Schematic diagram to illustrate how antibiotic gene clusters are controlled. The DNA encodes a number of genes clustered together on the chromosome. The genes (rectangles) are usually transcribed as polycistronic mRNAs (shown by the unbroken straight arrows). Transcription of the production genes (unshaded rectangles) must be dependent on transcription of the positive activator (master gene, shown by the shaded rectangle and the protein by the two circles) that migrates and binds to the DNA (curved arrows) to allow transcription of the production genes to take place. Without the activator protein, there is no expression of production genes.

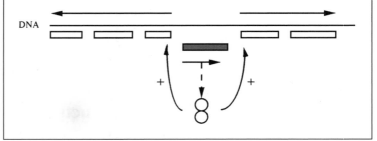

MASTER GENE
(POSITIVE
ACTIVATOR)
A positive activator is a
protein that is absolutely
required to bind to the
DNA upstream of the
target gene to allow it to
be transcribed, i.e. the
target gene cannot be
expressed without it. If a
collection of genes has
the same binding
sequence upstream, then
they are switched on
simultaneously. Hence
the positive activator acts
as a 'master switch' that
allows the entire set of
genes to be regulated co-
ordinately.

In this way, the cell ensures that the full complement of enzymes necessary to make the antibiotic are produced at the same
time, at the correct levels.

By cloning this master gene regulator and then expressing it
at unnaturally high levels, the cellular complement of the entire
biosynthetic machinery for production of an antibiotic may be
enhanced. Of course, there has to be sufficient fuel (metabolic
precursors and energy) to realize the full potential available from
this boosted level of biosynthetic machinery. The roles of metabolic flux analysis (Chapter 6) and process optimization in assur-
ing this advantage are very important.

It is also possible to force expression of the master gene during the growth phase and so to produce antibiotics during growth.
In some instances, this may be to the advantage of the overall
process, but often it is better to mimic the situation in Nature:
firstly to focus the design of the process on maximizing the acquisition of biomass, and then to turn that biomass to best advantage by using the cells as a factory (with the maximal level of
installed biosynthetic capability) to make the product at a fast
rate and to achieve a high titre. Therefore, controlled expression
of the master gene so that it is switched on decisively at the end
of growth is often preferred.

Molecular genetic analysis has also shown that some of the
strains from improvement programmes, isolated over the years
by random mutagenesis and selection, have increased dosages of
the biosynthetic genes. Thus, some of the best strains for penicillin production have multiple copies of the critical part of the
biosynthetic gene cluster arranged in tandem arrays. This effect
can also be achieved by gene cloning strategies.

The biochemical pathways for antibiotic biosynthesis are long
linear series of enzyme reactions. The flux through the entire
pathway is governed by the pace of the slowest catalytic step. If
the flow of metabolic intermediates through that step can be
improved, then there is a good chance that the productivity of
the overall process will be improved. Thus for the antibiotic
tylosin, produced by *Streptomyces fradiae*, it was established
that the rate of the last step in the pathway, conversion of
macrocin to tylosin by macrocin-O-methyltransferase (encoded
by the gene *tylF*) limited the overall productivity. As the strain
improvement programme had developed with time and the production strain lineage was reviewed, it was apparent that mutants with higher titres of tylosin also displayed extremely high
levels of macrocin which was excreted as a shunt metabolite
(Figure 5.9) because conversion of macrocin was limiting. Strains
which had the methyltransferase gene cloned and expressed at
high levels showed improved titres of tylosin.

Often, process analysis shows that some metabolites on the

Figure 5.9

Schematic diagram highlighting (a) the limitation of tylosin biosynthesis by *Streptomyces fradiae* and (b) the elevation of such limitation by the use of a recombinant strain.

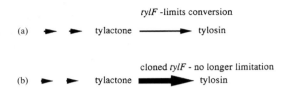

Figure 5.10

Schematic diagram highlighting the problem of shunt products and how this can be prevented. The metabolic pathway (substrate a to product d) is composed of three enzymes (1, 2, 3) with intermediates b and c. Because the carbon flow through steps 1 and 2 is greater than through step 3 (shown by the width of the arrows), the shunt metabolite, X, is formed through the action of enzyme 4. The presence of shunt metabolite, X, may complicate the recovery and purification of the desired product, d. By making a mutant devoid of enzyme 4, this is prevented.

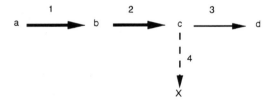

main biosynthetic pathway are being diverted to other shunt products (Figure 5.10). This not only represents a waste of valuable carbon source, but also presents a problem for the ultimate purification of the desired product, as the second metabolite has to be purified away. Genetic manipulation can be used to specifically inactivate the gene for the enzyme that diverts the metabolite to the shunt metabolite. This precise 'genetic surgery' enhances the flow through the pathway to the desired product.

One last important contribution that genetic manipulation makes is in providing flexibility of utilization of a carbon source. If, for example, there is a plentiful supply of cheap lactose available (from the milk whey industry) but the strain does not use lactose naturally, then lactose utilization genes can be cloned from another species and introduced into it.

At times, the world market-place has a glut of hydrolysed sucrose. This cheap carbon source is problematic for fermentation because one of the hydrolysis products (glucose) is capable of inhibiting the utilization of the other (fructose). It is possible to overcome this problem by the cloning of the fructose utilization genes which, in turn, ensures that the two monomeric carbon sources are used concurrently. The overall economics of such a process benefits from being able to use a cheap and plentiful carbon source in an efficient manner, without excess fructose carbon being left in the broth at the end of the fermentation.

5.10 Strain improvement for high-value recombinant products

Strain improvement programmes for products derived from cloned genes follow a completely different strategy from those undertaken for their lower value counterparts. The techniques used to clone a gene (or cDNA) usually place it precisely in a vector (most often, a plasmid), in a context that allows continuous high level expression or, more advisedly, expression of the gene is controlled and induced at a high level only after growth has ceased – in the same way as antibiotic fermentations are performed (Figure 5.2). A countless number of possibilities arise to tailor the gene within the vector, but this aspect involves only molecular biology and falls outside the remit of this chapter. The biology of the host strain is an equally critical factor in the overall performance of a fermentation process for a recombinant product, and strain improvement programmes can have a significant impact on the economics of such a process.

In the early days of the 'new biotechnology' industry, considerable difficulty was experienced in 'scaling up' recombinant processes from the laboratory to the pilot plant (i.e. from a scale of around 100 ml to 400 l). This was because, at the larger volume, a greater proportion of cells had lost the recombinant plasmid from the cells. Plasmid-deficient cells do not make the recombinant protein, and this reduces the overall productivity. The root cause of the problem is the number of cell generations needed to attain significant growth of the culture at the larger volume, coupled to the inherent instability of engineered plasmids in cells. The growth rate of a cell carrying a plasmid is invariably slower than its counterpart that has shed its plasmid load. Therefore, plasmid-deficient cells will outgrow the plasmid-containing cells in a population, a phenomenon which becomes more significant as the number of generations in a culture increases. The larger the volume of the fully-grown culture, the greater the likely proportion of plasmid-deficient cells in its population.

Figure 5.11
Plasmid segregation – the complication of plasmid oligomers. In case (a) the plasmids are present as four monomers and two segregate (on average) into each of the daughter cells. By contrast, in case (b), the plasmids are present as a tetramer. During growth, one daughter cell receives the tetramer, whereas the other does not – and becomes plasmid-deficient.

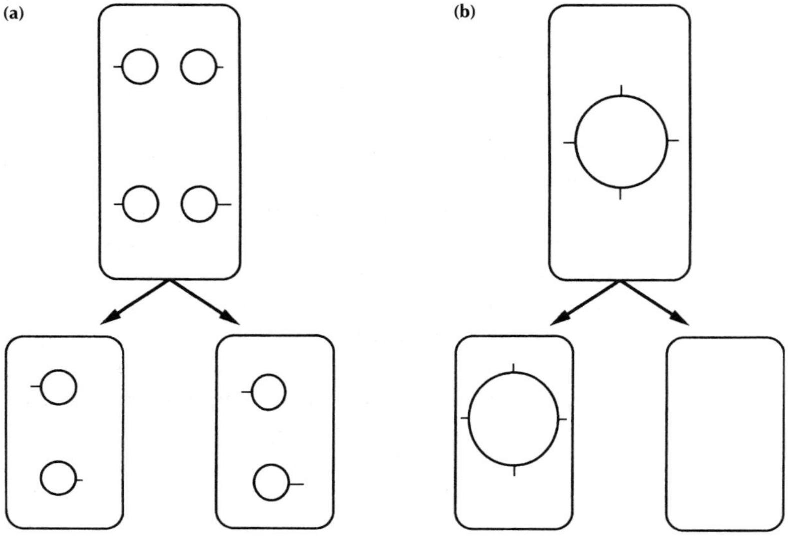

Resistance to an antibiotic, encoded by a plasmid-borne gene, is the usual selection strategy for the presence of a plasmid in a recombinant culture. The most commonly used selection is resistance to ampicillin, encoded by β-lactamase. Ampicillin-resistant cells survive because they are protected by the β-lactamase, which breaks down the chemical backbone of the antibiotic. Eventually, all of the ampicillin in the fermentation broth is broken down and the selective pressure is lost – the longer the fermentation (i.e. the more generations that take place), the more likely it will be that the antibiotic will become inactivated and that plasmid-deficient cells will form.

The frequency at which plasmid-deficient cells are formed, in the absence of antibiotic, is biased because of their natural tendency to form oligomers within the cell. Consider a theoretical situation in which a cell has a plasmid copy number of four. This may be conceptualized as four separate plasmids, two of which segregate into each daughter cell at division (Figure 5.11a). However, because of oligomerization, an equally common situation is that there will be a single tetramer of plasmids within the parent cell (i.e. the copy number will still be four). When the daughter cells are formed, one will receive the plasmid, which consists of four monomers, and the other will not (Figure 5.11b). In the absence of selective pressure (i.e. after all of the antibiotic has been exhausted) both will survive, and the plasmid-deficient daughter will outgrow its plasmid-containing sibling because of the advantage in growth rate.

Engineered plasmids undergo such oligomerization events at alarming rates. However, natural plasmids do not undergo such 'segregational instability', because they have a natural tendency to break back down from oligomers to monomers again. This ability has been lost during the process of conversion of natural plasmids into genetically-engineered vectors. It was discovered that a small piece of DNA, called *cer*, was the missing factor in the engineered vectors. When *cer* was reintroduced into vectors from the natural plasmids, the segregational stability of the plasmid construct containing the cloned gene to be expressed was improved. However, operation of *cer* is dependent on the strain and it is this aspect of the biology of plasmid instability that can be addressed by strain improvement programmes. The cellular machinery which allows *cer* to operate is rather complex and involves at least three different genetic loci. If any of the three is defective or missing, the result is that plasmids (even those containing a fully-competent *cer*) are unstable. Strain improvement of such cell lineages is best achieved using a targeted genetic approach, i.e. to use the power of genetic recombination to introduce the relevant machinery back into the chromosome of the strain, an extremely focused and targeted task.

Recombinant cultures are very prone to infection with bacterial viruses called **bacteriophages**. At a scale of fermentation above a litre, in a production environment, the recombinant cultures are at risk. There are two strain improvement strategies that can be brought into play. Firstly, it is possible to mutate the strain to become resistant to viral infection. This approach works for known viruses, but there is always the risk that infection from a new source will take place. A second, more secure strategy is to introduce a restriction/modification system into the cell line. The modification enzyme will alter the host's DNA so that it is no longer a target for the resident restriction system. However, viral DNA which enters the cell will be recognized by the restriction system and degraded, thus preventing the infection. The strategy can be enhanced further by introduction of a second restriction/modification system as a 'back up', should some of the viral DNA escape restriction by the first enzyme system. Again this is a very targeted approach to strain improvement.

Hosts for foreign gene expression often recognize the foreign protein as 'not natural' and selectively degrade it, thus reducing the overall productivity. The enzyme system which undertakes this function, the *lon* protease, works in a similar way in ordinary cells, selectively degrading ordinary cellular proteins which have been made defectively with altered amino acids. By knocking out the *lon* protease gene, degradation of the foreign protein is prevented, albeit to the detriment of the host cell which now has a slight growth rate disadvantage. Methods which generate *lon*-deficient cell lines are well established and extremely targeted.

In each of the examples in this section, there has been no need to use random mutagenesis and screening as the blueprint to derive the improved strain is direct and straightforward. Subsequent to this phase, strain improvement programmes use the random approach to 'fine tune' the genetic content of the production recombinant strain to gain further advantages in productivity. Thus, the order of events is the reverse of that for the lower value counterparts.

Summary

Fermentation is an important source of many clinically useful drugs, and this trend is likely to increase in the future. The microbiologist, in collaboration with other scientists, has an important role to play in generating new microbial strains that produce higher levels of the desired metabolites. There is an impressive array of techniques that can be used to achieve this goal, from random mutagenesis to the latest advances in genetic manipulation. Each has a role to play in the overall process, but the priorities of using different approaches will depend on the nature of the fermentation and the product being made.

Suggested reading

Baltz, R.H. (1998) Genetic manipulation of antibiotic-producing *Streptomyces, Trends in Microbiol.*, **6**, 76–83.

Chater, K.F. (1998) Taking a genetic scalpel to the *Streptomyces* colony, *Microbiology*, **144**, 727–738.

Griffiths, A.J.F. *et al.* (1996). *An Introduction to Genetic Analysis*, 6th edn, New York: W.H. Freeman & Co., pp. 592–600.

Hardy, K.G. and Oliver, S.G. (1985) Conventional strain improvement, protoplast fusion and cloning, in Higgins, I.J., Best, D.J. and Jones, J. (eds) *Biotechnology: Principles and Applications*, Oxford: Blackwell, pp. 257–282.

Hockney, R.C. (1994) Recent developments in heterologous protein production in *Escherichia coli, Trends in Biotechnol.*, **12**, 456–463.

Segura, D., Santana, C., Gosh, R., Escalante, L. and Sanchez, S. (1997) Anthracyclines: isolation of overproducing strains by the selection and genetic recombination of putative regulatory mutants of *Streptomyces peucetius* var. *caesius, Appl. Microbiol. Biotechnol.*, **48**, 615–620.

Summers, D.K. (1991) The kinetics of plasmid loss, *Trends Biotechnol.*, **9**, 273–278.

Summers, D.K. and Sherratt, D.J. (1985) Bacterial plasmid stability, *Bioessays*, **2**, 209–211.

6 Optimization of Fermentation Processes by Quantitative Analysis: From Analytical Biochemistry to Chemical Engineering

David M. Mousdale, Jill C. Melville and Monika Fischer

6.1 Microbial fermentation as a chemical process

The use of microbial fermentations for the production of 'bulk fine chemicals' is a well-established technology (Figure 6.1) and began in the early 1940s with the development of large-scale processes for the industrial production of products as diverse as organic acids, enzymes and antibiotics.

Since the pioneering work with penicillins, many more antibiotics and related secondary product processes have emerged and been commercialized. Microbial fermentations have also been developed for the production of a wide range of pharmaceutical products and enzymes for use in the food and detergent industries. No accurate figures are generally available for the scale of production of these products, but it is certain that microbial fermentations play an important role in contemporary manufacturing practice; while the expression of 'foreign' proteins with biomedical importance has been achieved in sheep (and the products extracted from their milk), it is unlikely that 100 000 litre fermentors will be rapidly replaced by flocks of cloned sheep as 'factories' for the production of such potent biological molecules.

Today there are new demands and emerging horizons for the rapid development of new processes to a commercial scale. These include the widespread emergence of microbial resistance to older groups of antibiotics, the interest in quite novel biological activities in the microbial fauna of exotic natural sources (such as

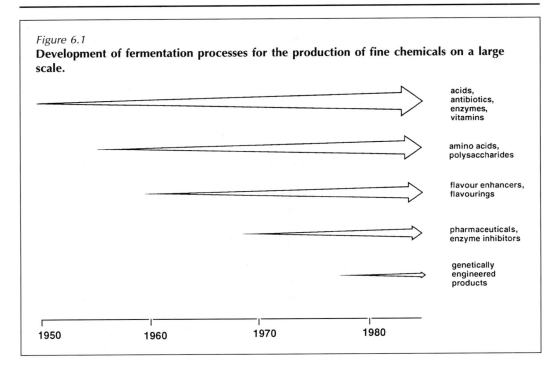

Figure 6.1
Development of fermentation processes for the production of fine chemicals on a large scale.

deep sea sediments, tropical rain forests and high temperature geological sites) and the environmentally (and economically) driven interest in using microbes to produce aromatics and other chemical building blocks from the renewable resource of photosynthetically-derived sugars.

Fermentation processes were initially devised and improved without the aid of modern analytical methods and techniques, mostly by random selection of mutagenized strains and short-term empirical improvements to nutrient media, feeding strategies, fermentor design and process operation. Analytical (bio)chemistry has made great advances in the last 50 years and it is now routine for laboratories to have ready access to powerful analytical facilities such as:

- high performance liquid chromatography (HPLC)
- capillary gas chromatography (GC)
- combined chromatography/mass spectrometry (LC-MS, GC-MS)
- ion chromatography (IC)
- ion-selective electrodes
- enzymatic assays for carbohydrates, carboxylic acids, etc.
- capillary electrophoresis (CE)
- supercritical fluid extraction and chromatography (SFE/SFC)
- two-dimensional polyacrylamide gel electrophoresis (2D-PAGE).

The advent of such techniques provides an alternative to the earlier 'black box' technologies where only the *results* of changes are assessed without any basis of understanding of the *causes*. In this chapter, a general method is outlined for the optimization of fermentation processes using an analytical approach. The framework of this method consists of the following stages:

- accurate quantitative analysis of the important chemical and biochemical events which are related to growth and product formation;
- identifying target areas for process improvement;
- devising practical means to change these key fermentation events;
- understanding the metabolic consequences of these changes.

Although biological systems are complex, that complexity is not a barrier to practical intervention. Biochemical teaching laboratories display impressive metabolic charts which include most, if not all, of the major biosynthetic pathways, with all the multiplicity of catabolic and anabolic pathways and their interactions. Direct analysis of what *actually* happens and, from a knowledge of the overall pathways involved, what *must* occur identifies the key targets for rational intervention and indicates which areas of the biochemical map require further investigation. This view of fermentation processes is derived from systems analysis: inputs (nutrients) and outputs (products) are linked by the generation and functioning of the biomass (the microbial cells). The focus is, therefore, on the process as a series of major events rather than on its fine details (Figure 6.2). Process

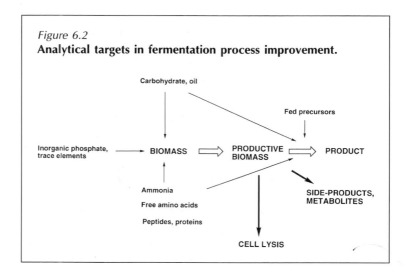

Figure 6.2
Analytical targets in fermentation process improvement.

optimization proceeds from what can be measured and does not require a complete understanding of either the molecular biology of the organism or the biosynthetic pathway of the product.

6.2   The utilization of fermentation inputs

The details of industrial manufacturing processes are confidential. For some processes, however, a wealth of practical information has been made available; for the purposes of illustration, therefore, reference will be made, where appropriate, to penicillin processes using data from or quoted in the sources listed at the end of this chapter. Penicillin is unusual in that it has a comparatively simple and well-defined biosynthetic pathway (see Figure 6.3); in most other respects, it is a good example of the older generation of antibiotic-producing fermentations. Progress curves showing input utilization, growth and product accumulation are computer simulations which exemplify those found in different fermentation processes at scales of production ranging from shake flask cultures to 100 m^3 vessels; the length of the hypothetical idealized process (7 days) is chosen to be

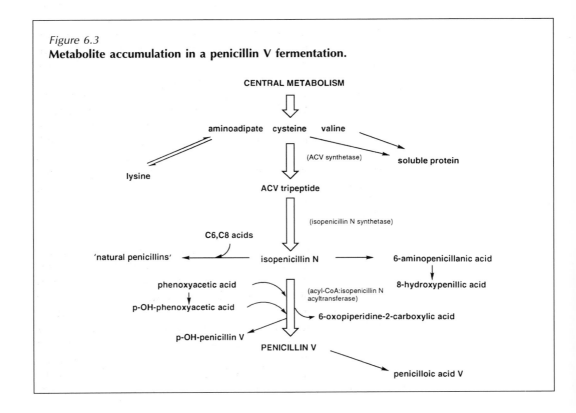

Figure 6.3
Metabolite accumulation in a penicillin V fermentation.

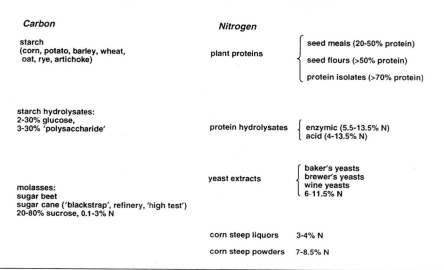

Figure 6.4
'Complex' carbon and nitrogen sources for industrial fermentations.

Carbon	Nitrogen	
starch (corn, potato, barley, wheat, oat, rye, artichoke)	plant proteins	seed meals (20-50% protein) seed flours (>50% protein) protein isolates (>70% protein)
starch hydrolysates: 2-30% glucose, 3-30% 'polysaccharide'	protein hydrolysates	enzymic (5.5-13.5% N) acid (4-13.5% N)
molasses: sugar beet sugar cane ('blackstrap', refinery, 'high test') 20-80% sucrose, 0.1-3% N	yeast extracts	baker's yeasts brewer's yeasts wine yeasts 6-11.5% N
	corn steep liquors	3-4% N
	corn steep powders	7-8.5% N

intermediate between the shorter fermentations with bacterial species and the lengthier processes using yeasts and fungi.

On an industrial manufacturing scale it is seldom economical to use chemically defined media (for example, a sugar carbon source, an ammonium salt as the nitrogen source, and a few other inorganic compounds such as a 'trace elements' mix). Inexpensive but complex carbon and nitrogen inputs are commonly used which are themselves products or by-products of other industrial processes (Figure 6.4). More than one carbon and nitrogen source is often required for optimum product yield. To some extent, this can be rationalized as different carbon sources (for example, a carbohydrate and oil) feeding into the major pathways of metabolism at different points; protein, peptides and free amino acid inputs are also utilized and these could differentially support either growth or product formation.

In addition to carbon and nitrogen, the chemical inputs also include phosphate and other inorganic ions and trace elements (iron, zinc, cobalt, molybdenum, etc.). The necessity for so complex a mix of nutrients can be assessed using this analytical approach which will establish the contribution of each additive to biomass and product formation; this allows the medium to be simplified and optimized. The medium developed for an existing production process is the result of a (probably lengthy) series of empirical trials and it may contain some important nutrients which are poorly utilized and which could be replaced with immediate advantage. To this extent, the medium may, therefore, be the 'message', in the sense of the famous sound-bite from the 1960s. At the very least, the medium is the first focus for analytical scrutiny as heat sterilization may itself 'damage' some of the nutrients – the best known example is the Maillard reaction between sugars and amino compounds which generates large amounts of unwanted and probably unusable degradation products. The key ability required is that of accurately measuring the utilization of inputs, both quantitatively and as part of a temporal sequence of events initiated when the medium is inoculated with cells generated in the previous 'seed' stage.

6.2.1 Carbon sources

6.2.1.1 Carbohydrates

'THE MEDIUM IS THE MESSAGE' (MARSHALL McLUHAN 'UNDERSTANDING MEDIA' [1964])
A typical production medium for a penicillin V process (H. Jorgensen *et al. Biotechnol. Bioeng.* **46**: 558–572, 1995): 3 g/l sucrose, 100 g/l cornsteep liquor, 1 g/l KH_2PO_4, 12 g/l $(NH_4)_2SO_4$, 0.06 g/l $CaCl_2.2H_2O$, 5.7 g/l phenoxyacetic acid

Glucose and other monosaccharides and simple disaccharides (such as sucrose and lactose) are readily determined by HPLC or by chemical and enzymic assays. Readily utilizable sugars such as glucose frequently exert 'carbon catabolite' inhibition of secondary product formation and, once the sugars in a batched medium are exhausted, must be carefully fed to avoid their accumulation. It is likely that glucose feeds will be routinely monitored on-line in the near future with amperometric sensors based on glucose oxidase.

Starch hydrolysates and polysaccharide inputs present much greater analytical difficulties. Straightforward test-tube chemical and enzymatic assays can measure the overall utilization of the carbohydrate (Figure 6.5). This type of carbohydrate input is, however, a population of molecular species with different molecular weights: in starch hydrolysates, for example, the average glucan oligomer may be maltoheptose, but different methods of manufacture can give variable mixtures of oligosaccharides. This is important because microorganisms utilize the different oligomers at quite different rates (Figure 6.5). Part of this utilization is enzymic, i.e. the breakdown of oligoglucans to either free glucose or (if the organism cannot take up glucose) maltose or maltotriose. For example, amyloglucosidase action liberates free glucose from glucans with α-1,4 and (at a slower rate) α-1,6 linkages, while α-amylase hydrolyses larger oligomers but only at α-1,4 linkages and pullulanase liberates large α-1,4 glucans by breaking α-1,6 linkages.

'DEGREE OF POLYMERIZATION' AS A SHORTHAND DESCRIPTION OF OLIGOSACCHARIDES
The degree of polymerization (dp) is a measure of the number of glucose units in a linear oligoglucan:
• glucose is dp1
• maltose is dp2
• maltotriose is dp3
....
• maltoheptose is dp7
etc.

Starch requires some pre-treatment before inclusion in nutrient media otherwise semi-solid preparations could result after heat sterilization. Amylase action breaks down the polymeric starch but only incompletely, leaving much of the carbohydrate in the form of large oligomers, some of which represent the 'limit dextrins' of classical carbohydrate chemistry and resist further enzymic attack at the α-1,6 branch points. Efficient starch

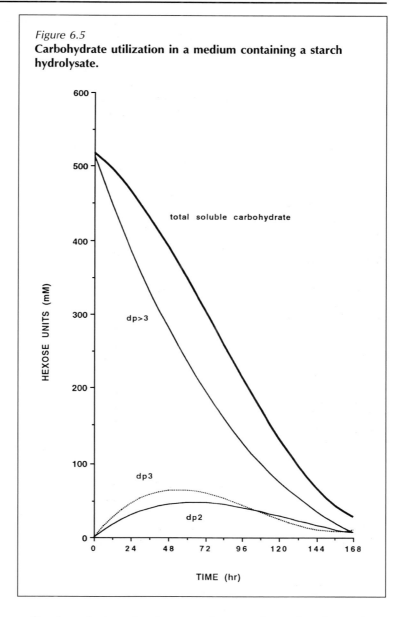

Figure 6.5
Carbohydrate utilization in a medium containing a starch hydrolysate.

utilization during the fermentation requires, therefore, three separate enzyme activities – amylase, amyloglucosidase and pullulanase – generated and secreted into the medium by the biomass. The biological entity in the fermentation is thus itself an important part of input processing before the substrates enter the cell.

Given the results of Figure 6.5, several questions would immediately arise with implications for process development and strain improvement:

- Is the initial accumulation of small oligosaccharides a sign of a limitation on growth by another input (most likely in the nitrogen sources)?
- Conversely, would a higher starting content of small oligosaccharides give faster growth in a fermentation with sufficient nitrogen?
- Would feeding a specific oligosaccharide (for example, maltose) maintain productivity?
- What carbohydrate 'processing' enzymes can the cells generate and do strains of differing productivities give different spectra of enzymes?

6.2.1.2   *Oils*

Plant-derived oils (from monoculture crops such as soybean) are cost-effective carbon sources available on a large scale; these also find uses as metabolizable anti-foaming agents. The mixtures of triglycerides in oils can be analysed by chromatographic (HPLC, GC, SFC or gel permeation) or enzymatic methods for the intact triglycerides or the fatty acids liberated from them by lipase action. Oils could be included in the category of complex carbon sources because their exact chemical compositions are variable and reflect their geographical source.

Lipid metabolism has the disadvantage that the bioenergetics of fatty acid breakdown are adverse in comparison with carbohydrate utilization; the oxygen requirements for fatty acid oxidation are very high and can result in oxygen depletion in the medium. Oleic acid is an alternative to oils but, because of the toxicity of long-chain fatty acids to many microbial species, must be fed and its levels measured accurately to modulate the feed and avoid the accumulation of the unused fatty acid in the fermentation.

6.2.1.3   *Other carbon sources*

Ethanol, some alkanes (chain length 12–18) and simple carboxylic acids such as acetic acid are readily utilized by some microorganisms. In penicillin processes where corn steep liquor is a major component of the medium, high levels of lactic acid are present (corn steep liquors are products of fermentations in which most of the carbohydrate is converted to lactic acid); the lactate is a minor carbon source but is known to be metabolized by *Penicillium chrysogenum*.

The supply of multiple carbon sources is common in manufacturing practice and both oil and carbohydrates can be metabolized at the same time. This facility to simultaneously utilize different carbon substrates may be of great value because there is a growing family of antibiotics with complex structures derived from clearly different areas of primary metabolism. Polyene macrolides and vancomycins, for example, have backbone structures derived from acetate units (best supplied as an oil or as acetic acid) to which aminated sugars (readily formed from glucose) are attached. Microbial biochemistry 'allows' both acetate- and glucose-derived precursors to be generated from either carbohydrate or lipid inputs but maximum productivity and yield might depend on a precise balance of the two types of carbon input. Analysis would show how the utilization of the inputs proceeds kinetically at different stages of the process.

6.2.2   Nitrogen sources

6.2.2.1   Proteins

As with oils, plant seed proteins are major fermentation inputs on the industrial scale (in fact, the proteins themselves are crude by-products of the extraction of the more valuable oil from the seeds). Protein inputs are supplied in four forms:

- macromolecular (undigested) in the starting medium
- digested by a protease in the starting medium
- undigested in a single or combined feed
- proteolytically digested in a single or combined feed.

When used, enzymic digestion may proceed only as far as the generation of peptides of variable molecular weight. Feeds of intact or digested proteins may be with carbon or other nitrogen sources and this is usually determined by the requirement either to maintain optimum C:N ratios or to avoid providing substrates which might encourage excessive growth once product formation has commenced.

Protein utilization is a deceptively complex topic for analysis. The breakdown of protein during the course of the fermentation is easily followed by a standard macromolecular protein assay (Figure 6.6). Soluble peptides and free amino acids may, however, accumulate and their utilization might be kinetically quite different from that of a protein (Figure 6.7). Results such as those contained in Figures 6.6 and 6.7 suggest various lines for more detailed investigation:

- Soluble protein was readily broken down to peptides and free amino acids but how much of the protein in the starting medium was insoluble and how much of this insoluble protein became soluble during the process?
- The end of protein utilization just after 70 h was a crucial event in the process – what happened then?

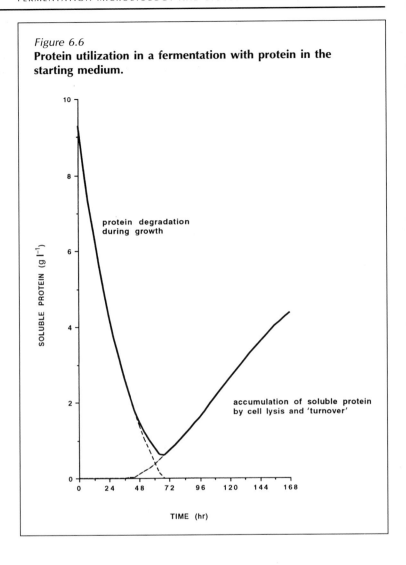

Figure 6.6
Protein utilization in a fermentation with protein in the starting medium.

- If the increase in soluble protein, peptides and free amino acids in the second half of the fermentation was caused by cell lysis, was this a consequence of the protein in the starting medium being fully utilized?

The extent of the utilization of a protein in a fermentation is usually measured only indirectly. If the biomass can be grown on a medium lacking protein inputs, the protein level in the cells expressed per unit biomass represents a baseline with which the complex medium fermentation can be compared. Protein contents in the biomass (in practice, the sedimentable material) approaching that found in the 'pure' biomass indicate complete utilization of that portion of the protein input which was insoluble,

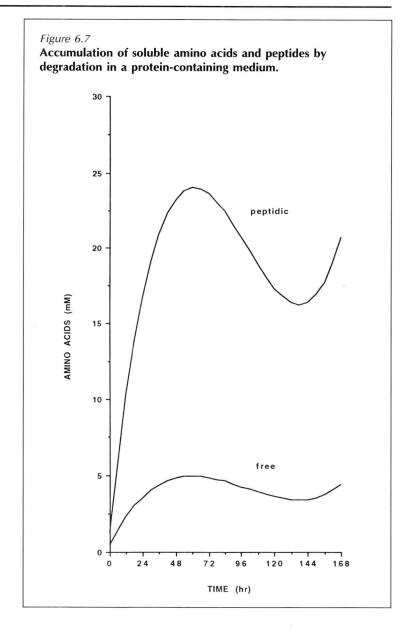

Figure 6.7
Accumulation of soluble amino acids and peptides by degradation in a protein-containing medium.

while persistently higher values suggest that the protein input is never fully used or, at least, solubilized (Figure 6.8). The varying degrees of protein utilization presented in Figure 6.8 could simply reflect the growth rates in the two fermentations, but the abilities of different strains to utilize macromolecular proteins and differences in the utilization rates of grades of any protein source would yield similar results.

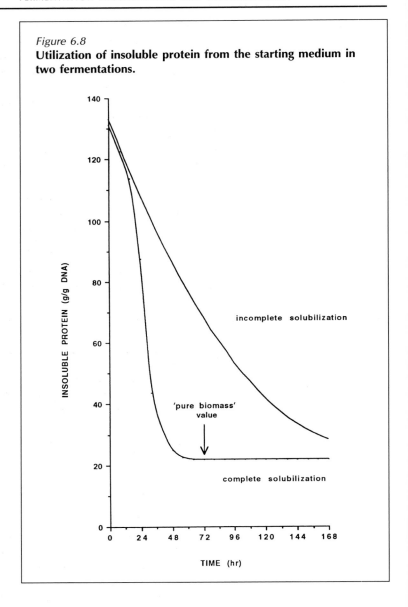

Figure 6.8
Utilization of insoluble protein from the starting medium in two fermentations.

6.2.2.2   *Peptides and free amino acids*

'Peptidic' nitrogen sources such as yeast extracts and corn steep liquors (see Figure 6.4) are heterogeneous in that they also contain some macromolecular protein and also low levels of free amino acids (only some casein digests contain most of the nitrogen as free amino acids without any macromolecular protein). Peptidic nitrogen sources are preferable to proteins as feeds because their amino nitrogen is already in a biologically more accessible form. Their quantitative determination requires acid hydrolysis of samples from which the proteins have been precipitated followed by correction of the total amino acid content for the (separately measured) free amino acids (Figure 6.7).

Free amino acids can be used as fermentation inputs but because of their cost are more often supplied in nitrogen sources such as yeast extract and corn steep liquor. The contents of the individual free amino acids are in these, however, quite variable and some may be present in very low concentrations. This is especially true for glutamine (which may have been cyclized to pyroglutamic acid) and the aromatic amino acids phenylalanine, tyrosine and tryptophan. Non-protein amino compounds are also present, some at quite high levels in corn steep liquors (ornithine, 4-aminobutyric acid, β-alanine, amino-alcohols, etc.).

When free amino acids are present in the starting medium, their rates of utilization are highly variable; some may even be exhausted while others are still present in readily detectable amounts (Figure 6.9). This is mostly due to the mismatching of the amino acids which are supplied to the inoculum with those required for the generation of new cells (this is a general feature of nitrogen sources for microbial growth and applies equally well to peptides and proteins over the course of the fermentation). Some amino acids are accumulated inside the biomass rather than simply polymerized into proteins (for example, proline and glutamate have roles in osmoprotection, and both glutamate and glutamine are required for biosynthetic amination reactions), and the rate of their disappearance from the extracellular phase need not correlate with utilization rate – this is a major inaccuracy in attempts to measure metabolic flux rates from analytical data obtained solely with culture filtrates.

6.2.2.3 Ammonia

Ammonia is the most important nitrogen source for secondary product formation because most of the nitrogen incorporated into organic compounds occurs by amination reactions using glutamate or glutamine as amino donors. Ammonia may be included in the starting medium in the form of ammonium salts (such as ammonium sulphate) and in complex nitrogen sources, and some free ammonia is liberated by the partial degradation of the complex nitrogen sources during heat sterilization. Ammonium salts may also be fed or gaseous ammonia supplied as a buffering agent when the accumulation of acids (or the uptake of ammonium ion in the starting medium) acts to depress the pH.

Industrial microorganisms vary greatly in their tolerance to high ammonia concentrations, and the monitoring of the ammonia level during the course of fermentation is very important. Figure 6.10 shows two contrasting free ammonia profiles. In

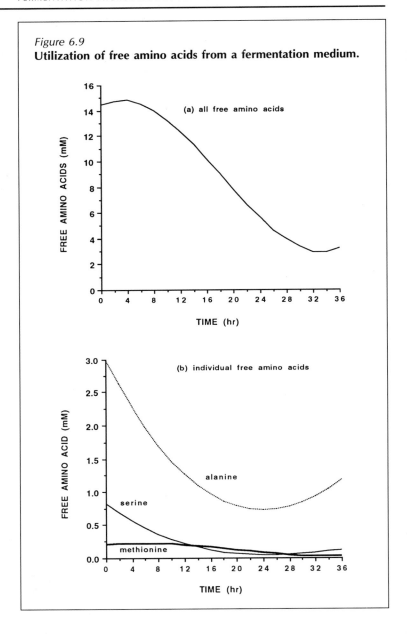

Figure 6.9

Utilization of free amino acids from a fermentation medium.

the first, the level of free ammonia rises as amino acids are deaminated and the carbon skeletons used for biosynthesis and/or catabolism; later the level of accumulated ammonia declines as ammonia is taken up by the growing biomass; in the third stage gaseous ammonia is fed in response to pH and the ammonia level rises greatly as carboxylic acids are formed. The second case shows the rapid utilization of the ammonia present in the starting medium, after which the level of free ammonia remains

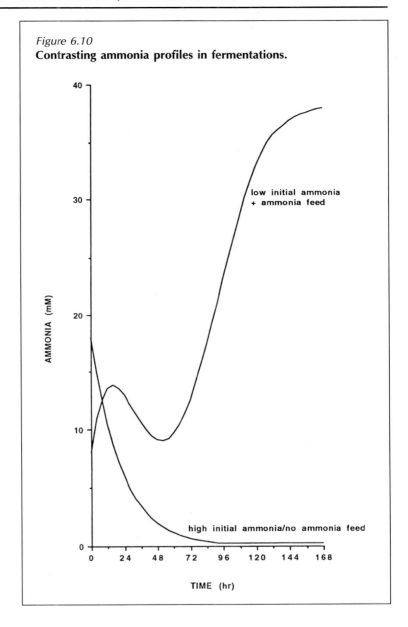

Figure 6.10
Contrasting ammonia profiles in fermentations.

low as no further ammonia or ammonium salts are supplied while organic nitrogen sources are used for growth and product formation. Ammonia feeding to the second process may, therefore, be a means of supplying a readily used nitrogen source which would sustain faster rates of product formation. Conversely, in the first process the trend towards higher product formation in improved strains risks generating higher ammonia levels (especially if the ammonia is balancing acid accumulation) which may eventually tax the tolerance of the cells to high ammonia concentrations.

6.2.3 Inorganic components

6.2.3.1 Macro-inorganics

Ammonia (as ammonium ion) is the most important inorganic cation because of its central role in nitrogen metabolism. Sodium, potassium, magnesium and calcium play different roles in fermentation processes. Sodium may be added as sodium hydroxide to adjust the pH of the initial medium or to control pH after inoculation; it is generally thought to be inert but may add to the osmotic potential of the medium. Potassium salts are used as sources of sulphate and chloride; potassium is accumulated by some microbial species as an osmoprotectant. Magnesium salts are also used as sources of chloride and sulphate while calcium, as calcium carbonate (chalk), is added to increase the pH of the medium; both of these cations participate in precipitation reactions of, for example, ammonia (as magnesium ammonium phosphate) and phosphate (as di- and tri-basic calcium phosphates).

Gaseous ammonia is NH_3. On dissolving in water, ammonium hydroxide (NH_4OH) is formed. The acid dissociation constant, pKa, of the ionic equilibrium is 9.25 at 25°C. The relative amounts of NH_3 and NH_4OH in a solution are determined by the pH:

- at pH 9.25 50% is in the form of NH_3
- at pH 8.25 only 10% is present as NH_3
- at pH 10.25 90% is present as NH_3

The pH in many fermentations is controlled to 6.0–7.5; in such culture media, therefore, free NH_3 is present in very small amounts.

Phosphate is the most important inorganic anion because it is essential for biomass formation (DNA, RNA, phospholipids, etc.). Phosphate levels in secondary product fermentations have received much attention due to the inhibitory effect of high phosphate concentrations on product formation. This is usually an indirect effect resulting from the stimulation of growth where rapid growth and high product formation are incompatible (this is discussed later when growth in fermentation processes is considered). Figure 6.11 shows an example of a process with low phosphate content in the medium and where that phosphate was rapidly utilized; such a fermentation may become phosphate-limited at a relatively early stage. This is in contrast to the other examples where much higher levels of phosphate are either incompletely utilized (and where the phosphate concentration is likely to be an important factor in pH regulation during the fermentation) or where the phosphate concentration only became low at an advanced stage in the process. In some cases, phosphate levels must remain low for high productivity (this is often the case with antibiotic-producing processes). Organic phosphates such as phytic acid which are introduced into the medium in complex nitrogen sources are potential sources of inorganic phosphate liberated by enzymic or chemical breakdown; these can greatly affect the phosphate supply during the fermentation.

The unequal utilization of the acidic and basic ionic species of salts has implications for pH control and regulation. The

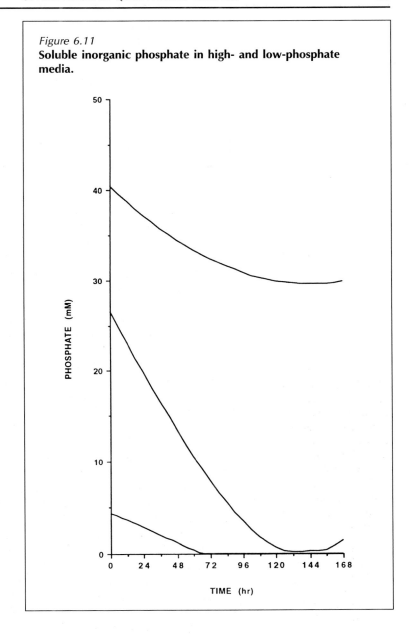

Figure 6.11
Soluble inorganic phosphate in high- and low-phosphate media.

uptake of ammonia from ammonium sulphate without the parallel uptake of sulphate by the cells would tend to force the pH down because of ionic exchange across the cellular membranes (in effect, leaving sulphuric acid in solution); the uptake of phosphate from potassium phosphate without an equal uptake of potassium would, on the other hand, be a move towards higher pH (leaving potassium hydroxide in the medium). The situation is made more complicated by the subsequent utilization of the counter-ions, and defining the mechanisms causing pH change in complex media can be very difficult. As well as contributing to pH, both chloride and sulphate ions may be directly incorporated into antibiotic production: for example, chloride is an essential precursor of vancomycins, chlortetracycline and avilamycins while sulphate (after reduction to organic sulphur) is a precursor of the cysteine required for the biosynthesis of penicillins, cephalosporins and lincomycins.

THE COMPLEXITY OF pH REGULATION IN A FERMENTATION

Factors driving pH down:

- accumulation of carboxylic acids
- utilization of bases (ammonium and potassium ions)
- net utilization of amino acids with basic side chains (lysine, etc.)
- CO_2 dissolving to form carbonic acid (important in early growth)
- plasma membrane H^+-translocating ATPases (in yeasts)

Factors driving pH up:

- utilization of carboxylic acids, phosphate, sulphate, etc.
- accumulation of bases (ammonia from amino acid deamination, etc.)
- net utilization of amino acids with acidic side chains (glutamate, etc.)

6.2.3.2   *Trace elements*

Iron, zinc, copper, cobalt and molybdenum are often 'essential' in the sense that in their absence either growth or product formation will be sub-optimal. Industrial media contain them in the 'ash' fraction of complex nitrogen and carbon sources, and the unpurified water used in large industrial process is another significant source; supplementation by the addition of salts may be necessary.

Such di- and trivalent metals can, however, be predominantly insoluble in fermentation media as a result of precipitation reactions. Phytic acid, contained in corn steep liquor and some grades of plant proteins, forms insoluble complexes with calcium, magnesium and heavy metals. There may be, therefore, a limited bio-availability of trace elements, whether they were added deliberately or gratuitously.

6.2.3.3   *Feeding of biosynthetic precursors*

To maximize their efficiency of utilization, biosynthetic precursors for product formation may be fed rather than included in the starting medium. The phenylacetic acid precursor for penicillin G is unusual in that high concentrations are toxic; a phenylacetate salt is routinely fed to a batch fermentation and its levels in the broth must be continuously monitored to avoid accumulation of the unused acid.

6.3   Growth and biomass profiles

6.3.1   *Measurements of growth and cellular contents*

The quantitative measurement of growth during the course of a fermentation is probably the single most important parameter for process optimization and yet is the most difficult and the most controversial. Cell counting methods are slow and difficult to apply except with unicellular organisms. Dry weight estimates and methods based on light scattering are accurate only in simple defined media. More indirect methods have been widely used including viscosity (especially for filamentous organisms), and oxygen uptake and carbon dioxide evolution rates.

Three biochemical measures, DNA, RNA and total cell protein, are generally the most useful if complex media are to be investigated (Figure 6.12); apart from their intrinsic value in elucidating the amounts of inputs used for biomass formation, they can be used to calibrate indirect measurements of growth for routine use, where this is possible. DNA estimates (either by the classical Burton procedure with acid-hydrolysed DNA or by the more recent and sensitive intercalating dye methods with extracted intact DNA) are the most accurate measures of growth because there are less interferences from media components. This is very useful when investigating the early phases of fast-growing fermentations where plant protein inputs interfere with accurate determination of cell protein and the plant cell wall carbohydrates associated with plant protein sources exaggerate RNA measurements which rely on the chemical estimation of ribose. Total RNA determinations reflect the ribosomal RNA content rather than the smaller transfer RNA contribution and the yet smaller and more transient messenger RNA in the cell. Cellular protein reflects the gross structural composition and, if the utilization of unsolubilized protein inputs can be quantified, gives an estimate of biomass which can be correlated with those from the nucleic acids. Other chemical methods include biochemical markers for cell wall polymers (for example, diaminopimelic acid) and, in oil-less processes, the total lipid content of the biomass.

Growth profiles such as exemplified by Figure 6.12 yield valuable information, i.e. separating phases of fast and slow growth, indicating later phases of net biomass loss, and giving estimates of the specific growth rate and of its change during the course of the fermentation. These are crucial aspects of the fermentation and high-priority targets for process optimization. The end of the early phase of most rapid growth is signalled by the exhaustion of such limiting factors as inorganic phosphate or one (or more) amino acids with low abundance in the starting medium. The potential auxotrophic factors also relevant to this part of the growth profile include vitamins and vitamin precursors contributed by complex nitrogen and some complex carbon sources, trace metal ions and assorted 'growth factors' (for example, purines and pyrimidines) on which the strain may have become dependent during strain development. This is sometimes apparent if seasonal trends in productivity or fermentation parameters can be related to changes in the source of complex inputs.

A different aspect of growth is that of biomass *composition*. Expressed on a weight basis, the content of both the macromolecular and the low molecular weight components varies in response to growth rate. Because fermentation processes other

Figure 6.12
Growth profiles by nucleic acids and cell protein measurements.

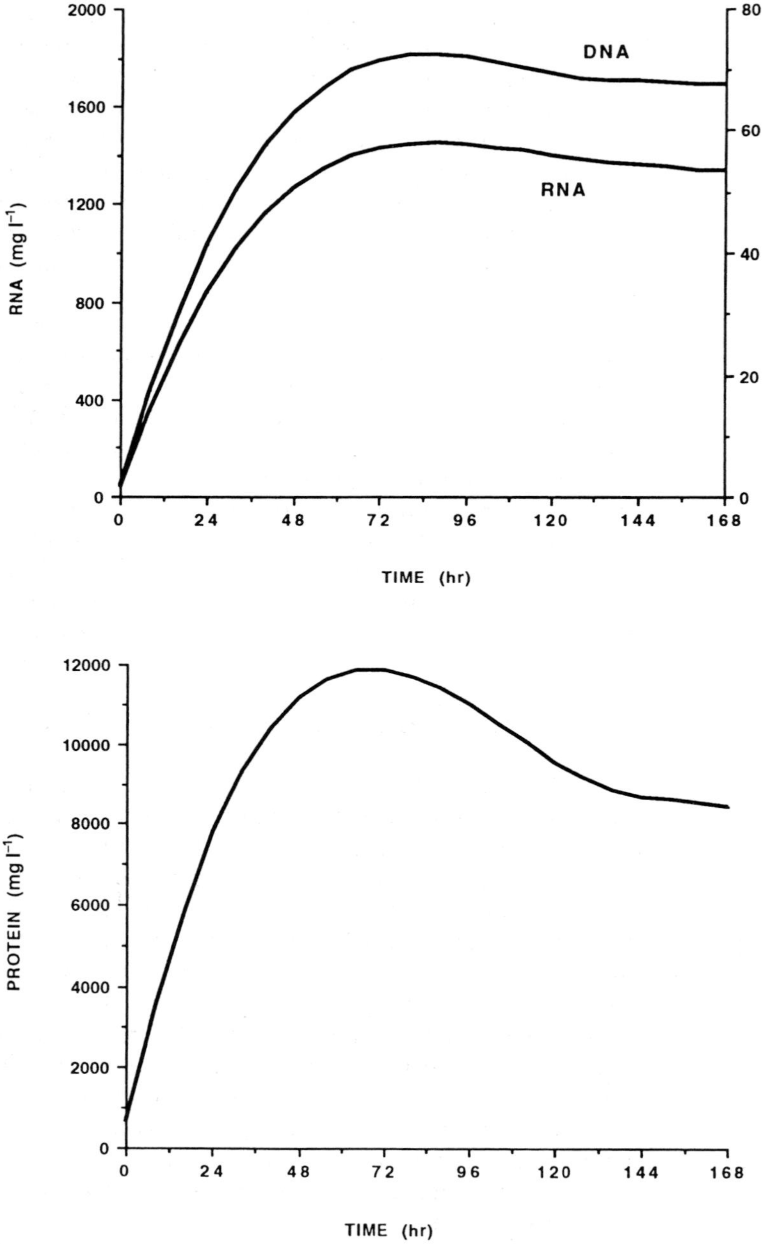

than in chemostats have a variable nutrient environment, there will also be an effect exerted by the availability of nutrients, but this is most important when the pool of low molecular weight constituents is considered. Unlike enteric bacteria, secondary product-forming prokaryotes (which are often filamentous) and eukaryotes have large pools of intracellular metabolites. This is, therefore, an important compartment inside the fermentor which is distinct from the soluble phase, and changes in the intracellular pools are sensitive indicators of the availability of soluble nutrients for biosynthesis. Phases of increasing intracellular pools indicate an excess of uptake over utilization while later phases of declining intracellular pools show a failure to maintain uptake (usually the exhaustion of the external solute) with respect to utilization.

Many secondary product-forming organisms exhibit a 'life cycle' in which a sequence of morphological changes can be associated with certain phases of growth and varying levels of productivity. These are reflected in major changes in DNA : RNA and DNA : total protein ratios and the levels of other cellular components, and these parameters are very useful in assessing the timing of transitions between phases and also the onset of contaminations. Published data on *Penicillium chrysogenum* show indeed that, at a faster growth rate in the early stages of a fed-batch process, both RNA and protein increase with respect to DNA but lipid and carbohydrate contents decrease.

6.3.2   *Growth and metabolism in the seed stage*

An indispensable but variable 'input' to a fermentation is the inoculum. All microbial fermentations are inoculated with the biomass which is formed after one or more previous 'seed' stages. Media used for seed vessels are frequently quite different from those used in the production vessel. This has many consequences, the most important of which is that the nutritional and environmental conditions experienced by the cells in the inoculum often cause a phase of adaptation after transfer to the production vessel. The metabolic pattern established in the seed stage will need to be 'unlearned' in the production vessel and the risk of adverse effects on initial growth rate and substrate utilization can be significant – for example, if a seed grown on glucose-containing medium is inoculated into a starch hydrolysate- or sucrose-based medium, the enzymes required for carbohydrate utilization are qualitatively different.

As with the production stage, growth in the seed stage is seldom measured directly and transfer criteria are consequently empirical (this adds a major uncertainty when process optimization by designed statistical trials of media or operational parameters is undertaken). Traditionally, the seed is optimized by using the production vessel, in effect, as a bioassay for the biological material in the seed tank. This is a seriously flawed methodology because the greatest influence on the outcome of the production stage may not be solely the metabolic competence of the cells in the inoculum. While a poorly grown seed is unlikely to show high productivity, a well-grown one may be adversely affected by operational parameters in the production stage. An accurate measure of biomass in the inoculum (by DNA, RNA, total protein, etc.) is essential; analysis of both what is and is not utilized by the time of transfer also defines potential problems in establishing rapid growth and development after transfer.

6.4   The accumulation of fermentation outputs

The accumulation of the product (such as an antibiotic) in an industrial fermentation is carefully monitored so that operational changes can be made to optimize yield in any one production-scale fermentation. This is frequently the only routine chemical assay employed other than pH and gas exchange (oxygen uptake and carbon dioxide evolution) once the fermentation has been inoculated. In terms of the carbon inputs to the process, the product may only represent 10% or less of the carbon supplied. In contrast, carbon dioxide can represent 50% or more, and accurate measurement of the carbon dioxide levels in the exit gas are essential if carbon mass balances are to be attempted.

There are, however, several classes of other outputs of quantitative importance, some of which may go unrecognized for many years: side-products, competitive products and accumulated primary metabolites. The biochemical relationships for the various types of metabolites accumulated in penicillin processes are summarized in Figure 6.3.

Side-products are either breakdown products of biosynthetic intermediates or of the final product itself (resulting from either chemical instability in the fermentation broth or from enzymatic breakdown). These two possibilities can be demonstrated during the course of penicillin production. The commercial product is either penicillin G (in which the aminoadipyl side chain has been exchanged for phenylacetic acid) or penicillin V (with phenoxyacetic acid). The aminoadipyl side chain of isopenicillin N (the final biosynthetic precursor for both penicillin G and V) partially accumulates after side chain exchange in a cyclized form, oxo-piperidine 2-carboxylic acid (first reported in 1980);

the efficiency of recycling of the aminoadipyl side chain may be variable in different fermentations. Similarly, the removal of the aminoadipyl side chain of isopenicillin N leaves 6-aminopenicillanic acid which appears in the extracellular medium where it is carboxylated to 8-hydroxypenillic acid. Penicillins are chemically unstable and the β-lactam ring can be opened to form penicilloic acids. Most biosynthetic pathways for secondary products are much more complex than the route to penicillins and are seldom fully defined; isolation of either major or minor side-products may, in these circumstances, be a useful aid to other biosynthetic studies.

Competitive products arise because the organism which was selected from its natural environment on account of one particular biological activity it exhibited may also have synthesized other natural products. Classical strain improvement can amplify as well as lose these other potential outputs, which are competitive in the sense that they represent a waste of nutrients and common substrates of central metabolism.

In the penicillin process there are two types of competitive products: firstly, the amino acid precursors of the penicillins can be used for other biosynthetic purposes; secondly, one of the enzymes of penicillin biosynthesis has a wide substrate selectivity. Cysteine and valine are common amino acids which are polymerized into proteins; once penicillin production has begun, therefore, *any* generation of soluble protein (including that of enzymes involved in carbohydrate and oil utilization) must compete with the use of the amino acids for penicillin biosynthesis. Similarly, α-aminoadipate is a biosynthetic precursor not only of penicillin but also of the amino acid lysine (in yeasts but not in bacteria); detailed work has shown that *Penicillium* strains with higher penicillin production have larger endogenous pools of α-aminoadipate and less of the precursor is used for lysine formation. Lysine can, under some circumstances, *stimulate* penicillin accumulation; this paradox was resolved when a catabolic pathway was identified which transforms exogenous lysine to α-aminoadipate. The acyltransferase which removes the aminoadipyl group from isopenicillin N and replaces this with a phenylacetyl or phenoxyacetyl group to form penicillin G or V, respectively, is the second source of competitive products. Ring-hydroxylation of phenoxyacetic acid produces p-hydroxyphenoxyacetic acid which can be incorporated by the acyltransferase into p-hydroxypenicillin V. The same acyltransferase will also accept several medium length (C6–C8) fatty acids to form other penicillins (the 'natural' penicillins DF, F and K). The use of oil in penicillin processes is much less common than glucose because triglyceride breakdown results in large amounts of fatty acyl derivatives and increases the levels of the natural penicillins.

Major products of primary metabolism include carboxylic acids such as citric acid which may accumulate as 'overspill' products from carbohydrate metabolism. The most important example of this is the formation of large amounts of acetic acid by recombinant strains of *Escherichia coli*; this has adverse effects on growth and the expression of foreign genes. Acetic acid and glycerol can also accumulate from the breakdown of triglycerides in oils. Free amino acids are formed from the proteolysis of protein and peptide inputs and, if in excess of the requirement for growth, can accumulate (Figure 6.7). A second category of primary metabolite is that of 'storage' products; many microorganisms accumulate, for example, the polysaccharide glycogen or the disaccharide trehalose, poly-hydroxybutyric acid, or phosphate polymers ('polyphosphates'), especially under conditions of nutrient limitation.

Enzymes generated and secreted by the biomass could be considered as 'primary' metabolites, but the absolute amounts of protein involved are usually small unless the organism has been selected or designed to produce large amounts of a particular enzyme (for example, the enzymes used in detergents and in the food processing industry). A certain amount of cell lysis ('turnover') occurs in fermentation processes and this has the result of releasing much larger amounts of cellular proteins into the soluble phase; soluble protein can, therefore, accumulate even though protein inputs may still be in the course of utilization (Figure 6.7). Similarly, soluble carbohydrate pools may increase despite no further carbohydrate being supplied as cell wall fragments accumulate from cell turnover during the fermentation.

Identifying side-products, competitive products and major metabolites is a key factor in process improvement. Their presence can be recognized by the application of class-selective chromatographic methods for:

- primary and secondary amino acids
- organic acids with UV detection at different wavelengths
- carbohydrates using refractive index and electrochemical detection
- reverse-phase chromatography
- isocratic and gradient ion chromatography.

Unexpected peaks in standard chromatograms, i.e. those not identifiable by reference to standard calibration mixtures, are the best sources of fermentation 'unknowns'. In general, the greater the accumulation of the unknown product the more likely it is to be of importance for process optimization. Once recognized, an unknown metabolite can be isolated by scale-up of conventional analytical HPLC procedures and structural identification undertaken by mass spectroscopy, nuclear magnetic resonance (NMR) spectroscopy and chemical synthesis.

6.5   Process improvement

The analytical results obtained from a fermentation form the basis for identifying the key events for intervention in process improvement. The first step is to relate the results to the kinetics of product formation.

In both batch and fed-batch fermentations product formation follows patterns exemplified in Figure 6.13. A constant rate of product formation can be achieved over lengthy periods of time in laboratory continuous fermentations but these have not yet been developed to the large production scale. The **yield rate** in a batch or fed-batch process invariably exhibits a maximum after which the product formation rate then declines.

There is a strong interaction between the rates of biomass and product formation. Historically, fermentations were divided into growth-associated and growth-independent; in the former active product formation occurs by actively growing cells, while in the latter the growth phase (trophophase) is followed by product formation under conditions of no net growth (idiophase). More recently, analysis has shown that these are extreme examples of a spectrum where the key parameters are the specific growth rate (μ), the specific productivity (q_{prod}) and the requirement for inputs to 'maintain' the biomass in the fermentor in a fully productive state. Mathematical simulations of increasing sophistication have postulated minimum and maximum μ values outside which maximum q_{prod} is not sustained and have made various assumptions as to how the carbon input is partitioned between growth, product formation and the provision of maintenance energy which is required by the biomass. The general outline of the productivity curves can be described as the generation of productive biomass followed by the loss of productive capacity as the number of competent cells declines; the process has an implicit life cycle in which young cells differentiate to productive cells which themselves only have a limited productive life, and it is the reduced rate of generation of the producing cells that underlies the failure to maintain the highest production rates.

Direct measurements of biomass (DNA, RNA, cell protein, lipid, etc.) will guide process improvement to achieve the optimal growth profile. This optimum is a combination of two factors: the fastest generation of productive biomass and the maximum length of time μ values can be maintained to allow maximum q_{prod}. The overall growth profile can then be optimized by the manipulation of fermentation inputs based on the kinetics of their utilization.

Which input limits the phase of rapid growth (at approximately 60 h in the idealized process of Figure 6.12)? The inorganic phosphate in the medium became exhausted at this time

---

DIFFERENT KINETIC ASPECTS OF PRODUCT FORMATION

The progress curves of product formation are generally sigmoidal. The tangent to the curve is an instantaneous production rate or **yield rate** (see Figure 6.13). Industrially, a more useful parameter is the amount of product formed per *accumulated* time after inoculation of the production fermentor (this is sometimes referred to as the **productivity**). The units of both are product mass (g, kg, tonne) per h.

The yield rate, and its profile during the fermentation, is related to changes in growth rate, utilization rates of inputs and metabolic events. The productivity measures overall kinetics of product formation and is important for estimating the economics of the process; factors such as the time required to empty, clean and re-fill the production vessel prior to the next inoculation can also be taken into account in the mathematics.

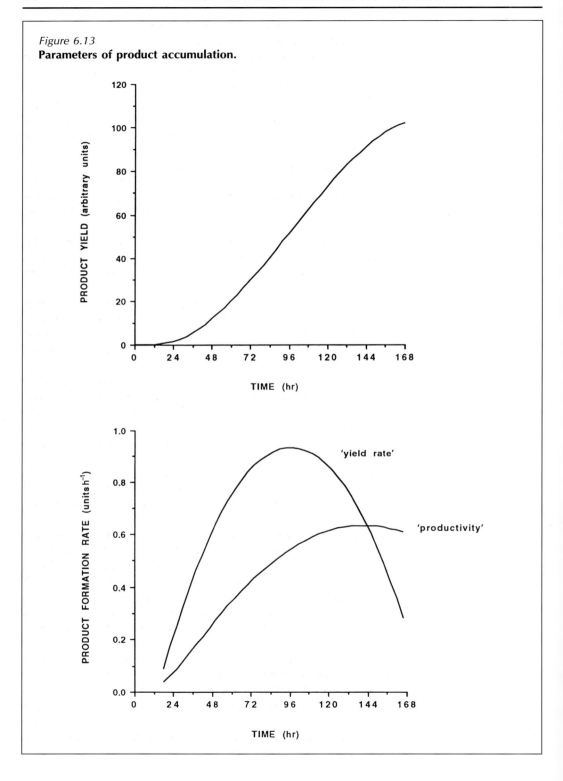

Figure 6.13
Parameters of product accumulation.

(lowest curve in Figure 6.11). The free ammonia may have been the nitrogen substrate for rapid growth (Figure 6.10, lower curve) or the protein in the starting medium might have fulfilled this role (Figure 6.6). Alternatively, the free amino acids may have been fully utilized (Figure 6.9).

Can the levels of the growth-limiting nutrient in the starting medium be increased to give a faster growth rate or higher biomass levels? Poorly growing fermentations are usually also poorly productive but cultures with high cell densities may become oxygen-limited. Simply increasing the concentrations may impose osmotic stress. A controlled feed may be applied once the nutrient in the batched medium becomes depleted, for example adding a phosphate feed to two of the processes illustrated in Figure 6.11.

Are there successive growth bursts (diauxic transitions)? This may be particularly important when 'rival' carbon sources such as carbohydrate/organic acids or sugars/triglycerides are present in the growth medium.

Can nutrient feeds sustain optimum μ values? The relationship between μ and specific productivity is well understood for penicillin processes and, given accurate growth data, the different stages of growth can be optimized by the choice of feeds and control of feed rates.

Do the macromolecular inputs require enzymic processing? The protein inputs in the example of Figure 6.7 were degraded during the process to soluble peptides and free amino acids but in other cases a protease digestion before the fermentation is inoculated may be necessary to make the nitrogen supply more biologically accessible.

Is the growth erratic? Irreproducible growth in a process can be caused by many factors, including variable precipitation (of phosphate, trace elements, etc.), inconsistent medium preparation and differing amounts of biomass in the inoculum.

Given an improved growth profile, the second key control point is the succession of factors which progressively limit product formation during the course of the fermentation. With a knowledge of the quantitative utilization of all the inputs which contribute to product formation, it is possible to compute the rates of utilization required to sustain the maximum rate of product formation; the first utilization rate to fall below that required for this maximum rate represents the first input (or substrate) limitation. This could be the major carbon source being consumed during product formation, a specific precursor either included in the batched medium or being fed, or nitrogen (as amino groups) required for amination reactions. The levels of the limiting substrate can, therefore, either be increased in the medium or fed (if it was only supplied in the batched medium) or its feed rate or timing could be optimized. The procedure is then repeated for each successive limitation, given the constraints of engineering, the volume in the fermentor and the osmotic strength of the starting medium. On a commercial scale, deciding the point in time at which higher time-based productivity could be gained simply by terminating the running fermentation (and setting up a new operating vessel) becomes the final criterion at any one stage in process improvement – in Figure 6.13 the likely harvest time would be 144 h.

Definition and measurement of side-products, unwanted and competitive metabolites leads to quite different options for process improvement. In general, higher substrate efficiencies follow the elimination of such undesirable products. Their elimination may, however, require gene technology or the careful scrutiny of strains developed by classical selection methods. The closer the source waste metabolite to the biosynthesis of the fermentation product, the larger may be the potential benefit to productivity.

Consider Figure 6.14 where the avenues and potential productivity gains by rational improvement of a hypothetical process are visualized as the partitioning of the supplied nitrogen into product, biomass and metabolites. Roughly equal amounts of the total input nitrogen are incorporated into the product and a side-product at the end of the fermentation; in a penicillin process this 'side-product' would represent the sum of the oxopiperidine carboxylic, aminopenicillanic, hydroxypenillic and penicilloic acids together with any natural penicillins formed. *If the accumulation of side-product could be eliminated, then the yield increase with this 'cleaner' process would be two-fold.* An obvious strategy would be to screen promising improved strains for those which accumulate less side-product and use these as a base for further strain development work.

With a competitive product the benefits of eliminating such a product are not as clear because the biosynthetic relationship with that of the main product must first be defined. In the case of penicillin V and p-hydroxypenicillin V, this relationship is very close; mutants could readily be obtained with much reduced levels of the hydroxylated product. Simplified product recovery during downstream processing would also be anticipated if the formation of a competitive product or side-product of similar chemical structure to the product were to be prevented.

A large amount of nitrogen remains unused in amino acids and peptides (Figure 6.14). This waste of unused nitrogen could represent peptides resistant to breakdown by the proteases secreted by the biomass (in which case a different mix of complex nitrogen inputs could be considered) or remnants of cellular proteins released by cell turnover. A related question is whether the biomass level in the process is unnecessarily high (i.e. of low specific productivity); could a different balance of inputs restrict growth and divert more nitrogen to product? These possibilities

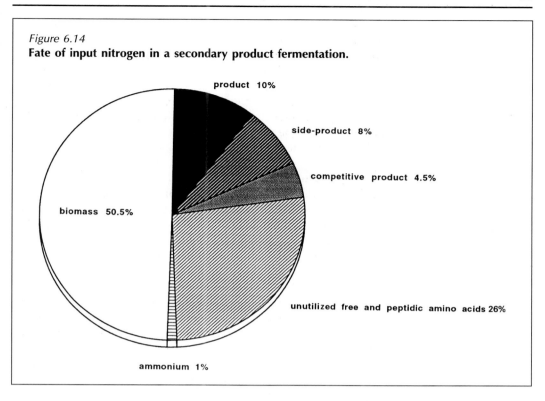

Figure 6.14
Fate of input nitrogen in a secondary product fermentation.

product 10%

side-product 8%

competitive product 4.5%

biomass 50.5%

unutilized free and peptidic amino acids 26%

ammonium 1%

should be explored at an early stage in process and strain development to avoid introducing problems which may become acute when large-scale fermentors are used. Scaling up from laboratory and pilot plant fermentors to production vessels may also result in differing quantitative patterns of metabolism and side-product formation which influence the success of the production process.

The information gained from growth studies, input utilization and output accumulation can be used to define targets for rational strain improvement. This can have a very large impact on productivity and does not require a complete knowledge of the genomics of the organism. This illustrates the central theme developed in this chapter: that direct analysis bridges the knowledge gap between the genetic potential of an organism and what portion of this is expressed in a given environment (Figure 6.15). By identifying what the cells *must* do (given a nutrient medium from which substrates for growth and product must be taken as building blocks) and what the cells *actually* do (accumulate side-products, competitive products, 'waste' metabolites, unused portions of inputs, etc.), the use of microbial cells as biocatalysts becomes amenable to precise genetic and metabolic engineering. Finally, the same approach can also be extended to cultures of

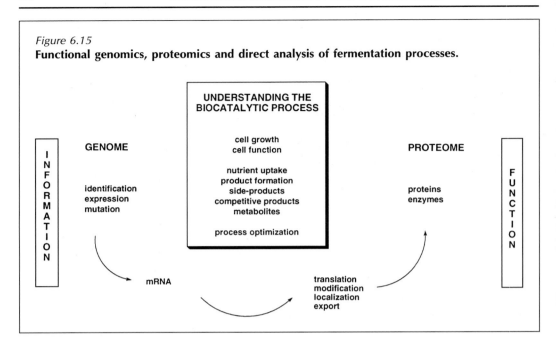

Figure 6.15
Functional genomics, proteomics and direct analysis of fermentation processes.

animal and plant cells with their diverse and mostly unexplored biosynthetic potentials.

6.6   Future prospects

Robotic sampling and analysis, on-line sensors for nutrient concentrations and cell density, 'instant' metabolite profiling by high-resolution mass spectroscopy and near-infra-red scanning methods, *in situ* NMR probes which can 'see' into cells and sub-cellular compartments, and artificial intelligence capable of re-acting to the data generated will revolutionize the operation of large-scale fermentors. Most of these technologies are already at an advanced stage of development. Biochemical analysis using two-dimensional gel electrophoresis methods will complement them with rapid analyses of the spectra of individual proteins made at various stages in the process and of how gene expression is quantitatively regulated. With that quality of information, process control will enter the *molecular* era and the full potential of the biological system represented by a fermentation can be achieved on a real-time basis.

Summary

Product yields in industrial fermentations have been increased by strain selection (both rational and random) and engineering (technological) improvements. Both approaches are limited by the inability to define the metabolic causes and consequences of the changes which are made to the biological system. Modern chemical and biochemical analytical techniques can now give a rapid understanding of the key events in a fermentation process: the utilization of inputs, growth, and the generation of product, side-products and unwanted metabolites. A detailed knowledge of input utilization, growth and the generation of outputs allows the optimization of growth profiles and the metabolic engineering of more advanced and efficient fermentation-based manufacturing. The benefits are that key targets can be selected for genetic manipulation and improved process control by optimization of the media, feeding strategies and physical parameters. Novel strains and processes can be more readily scaled up and the optimal features of strains with increased productivity can be defined more rapidly. This approach benefits both small- and large-scale processes, with prokaryotes and eukaryotes, and for the full range of products produced by contemporary fermentation biotechnology.

Further reading

Chang, L.T., McGrory, E.L., Elander, R.P. and Hook, D.J. (1991) Decreased production of *para*-hydroxypenicillin V in penicillin V fermentations, *J. Ind. Microbiol.*, 7, 175–180.

Esmahan, C., Alvarez, E., Montenegro, E. and Martin, J.F. (1994) Catabolism of lysine in *Penicillium chrysogenum* leads to formation of 2-aminoadipic acid, a precursor of penicillin biosynthesis, *Appl. Environ. Microbiol.*, 60, 1705–1710.

Greasham, R.L. and Herber, W.K. (1996) Design and optimization of growth media, in Rhodes, P.M. and Stanbury, P.F. (eds) *Applied Microbial Physiology. A Practical Approach*, Oxford: IRL Press (Oxford University Press), pp. 165–192.

Hersbach, G.J.M., van der Beek, C.P. and van Dijck, P.W.M. (1984) The penicillins: properties, biosynthesis, and fermentation, in Vandamme, E.J. (ed.) *Biotechnology of Industrial Antibiotics*, New York: Marcel Dekker Inc., pp. 46–140.

Hönlinger, C. and Kubicek, C. (1989) Metabolism and compartmentation of α-aminoadipic acid in penicillin-producing strains of *Penicillium chrysogenum*, *Biochim. Biophys. Acta*, 993, 204–211.

Jorgensen, H., Nielsen, J., Villadsen, J. and Mollgaard, H. (1995) Metabolic flux distributions in *Penicillium chrysogenum* during fed-batch cultivation, *Biotechnol. Bioeng.*, 46, 558–572.

Luengo, J.M. (1995) Enzymatic synthesis of hydrophobic penicillins, *J. Antibiotics*, **48**, 1195–1121.

Mousdale, D.M. (1996) The analytical chemistry of microbial cultures, in Rhodes, P.M. and Stanbury, P.F. (eds) *Applied Microbial Physiology. A Practical Approach*, Oxford: IRL Press (Oxford University Press), pp. 165–192.

Noronha Pissara, P. de, Nielsen, J. and Bazin, M.J. (1995) Pathway kinetics and metabolic control analysis of a high-yielding strain of *Penicillium chrysogenum* during fed batch cultivations, *Biotechnol. Bioeng.*, **46**, 558–572.

Paul, G.C. and Thomas, C.R. (1996) A structured model for hyphal differentiation and penicillin production using *Penicillium chrysogenum*, *Biotechnol. Bioeng.*, **51**, 558–572.

Stouthamer, A.H. (1978) Energy production, growth, and product formation by microorganisms, in Sebek, O.K. and Laskin, A.I. (eds) *Genetics of Industrial Microorganisms*, Washington, DC: American Society for Microbiology, pp. 70–76.

Tiller, V., Meyerhoff, J., Sziele, D., Schügerl, K. and Bellgardt, K.-H. (1994) Segregated mathematical model for the fed-batch cultivation of a high-producing strain of *Penicillium chrysogenum*, *J. Biotechnol.*, **34**, 119–131.

Zangirolami, T.C., Johansen, C.L. and Nielsen, J. (1997) Simulation of penicillin production in fed-batch cultivations using a morphologically structured model, *Biotechnol. Bioeng.*, **56**, 593–604.

7 Flux Control Analysis: Basic Principles and Industrial Applications

*E.M.T. El-Mansi and
Gregory Stephanopoulos*

7.1 Introduction: traditional versus modern concepts

Industrial microbiologists, biochemists and engineers are generally, but not entirely, of the view that flux through a given pathway is usually limited by one step. Such a step was termed the **rate-limiting step** or the 'bottleneck' with the enzyme catalysing such as a step being referred to as the 'pacemaker'. However, the question of how such a step can be identified and quantified in a given pathway remained unanswered, largely due to the lack of an experimental procedure which describes how such a parameter as 'rate-limiting' can be identified and quantified.

Clearly, if a rate-limiting step exists in a given pathway, then increasing the activity of the enzyme catalysing such a step will increase the overall flux through the pathway and, by the same token, varying the activity of any other enzyme will have no effect whatsoever on the overall flux of the pathway in question. Although a number of studies have been published in support of the concept of the rate-limiting step, the majority of studies which do not support it have not found their way into the public domain. However, a few studies have been published and the findings clearly indicate the inadequacy of such a concept. For example, a 3.5-fold (350%) increase in the activity of phosphofructokinase, the enzyme widely regarded as the rate-limiting step in glycolysis, had no appreciable effect on the flux through the glycolytic route of *Saccharomyces cerevisiae*.

By and large, the concept of the rate-limiting step does not appear to be a tenable proposition as it does not adequately explain why flux and, in turn, yield could not be improved following the overexpression of enzymes which are considered to be rate-limiting. For example, pioneering work on the production of lysine by *Corynebacterium glutamicum* by Stephanopoulos and Vallino (1991) identified PEP carboxylase as the rate-limiting

The **rate-limiting step** was defined as the slowest step in a given pathway. Such a definition resulted from the observation made in the mid-1960s that the reaction rate of a sequence of unsaturated enzymes (i.e. where the concentration of substrates is below the K_m value for each of the enzymes involved) depended non-linearly on the kinetic parameters of all the enzymes involved. However, no theoretical basis was given to validate or substantiate the existence of such a concept.

enzyme in this process. Although their conclusion was based on well founded reasons, further studies revealed that full inactivation of this enzyme had no effect on flux to lysine formation (Gubler *et al.*, 1994). Furthermore, the enzymes which catalyse rate-limiting steps tend to be subjected to regulation and any changes to their properties will inevitably affect the overall velocity of carbon flux through the pathway. A number of techniques have been devised to assess the relative contribution of each enzyme to the overall velocity of a given pathway and although *in vitro* measurement of the maximal velocity of a particular enzyme is useful, the value obtained does not necessarily reflect the rate of catalysis *in vivo*, due to a lack of hard information regarding the intracellular level of substrates and effectors, not to mention the rapid turn-over of metabolic pools which renders such assessment difficult if not impossible. A new quantitative approach was therefore called for, not only to explain the many outstanding observations relating to flux control in industrial fermentations but also to provide a rational basis for the exploitation of the diverse array of metabolic pathways.

Although a number of approaches have been used as an alternative to the rate-limiting step, including sensitivity analysis, an approach used to tackle similar problems in economics, the Biochemical Systems Theory (Savageau *et al.*, 1987), and the 'metabolite balance technique' used for the calculation of carbon flux to acetate excretion (El-Mansi and Holms, 1989), amino acid production (Vallino and Stephanopoulos, 1993) and in general (Holms, 1996), it is Kacser's theory of **Metabolic Control Analysis** (MCA) that has grown in stature since its inception in 1973 and proved, without undermining the intellectual capacity of other approaches, to be the ultimate approach. The controversy over the question of whether a given method can be used successfully to predict or determine the relative contribution of each enzyme to the overall flux in a given pathway was finally resolved when Kacser and Burns (1973) and Heinrich and Rapaport (1974) independently proposed the aforementioned theory of MCA. The fundamental difference between the rate-limiting step as a concept and that of the theory of MCA is that while the former is a qualitative parameter, the latter examines biological systems in a quantitative way that excludes bias, expectations or preconceived ideas.

7.2   Flux control analysis: basic principles

Kacser's MCA theory facilitates the assessment of not only how perturbation of a particular enzymic activity affects metabolic flux, but also by how much. The response to changes in the concentration of a particular enzyme on flux varies over a wide range. For example, the response could be immediate, with a strong correlation between the increase in flux and the increase in enzymic activity, as is the case for adenylate kinase *en route* to ATP synthesis. In this case one might justifiably describe such an enzyme or a step as rate-limiting. The majority of enzymes, however, do not enjoy such a high profile and as such an increase in flux may or may not be brought about by the increase in enzymic activity.

Furthermore, the degree of control exerted by a particular enzyme on the overall flux of a given pathway is not purely dependent on the numerical value of its intracellular concentration but rather on whether the enzyme has the capacity for higher throughput, which can only be ascertained from the value of the enzyme's **flux control coefficient**, the first pillar in the theory of MCA.

7.2.1   *The flux control coefficient*

The **flux control coefficient** is a parameter which describes in quantitative terms the relative contribution of a particular enzyme to **flux control** in a given pathway. It is not an intrinsic property of the enzyme *per se* but rather a **system property** and so is subject to change as the environment changes. It is generally expressed as the fractional change in flux in response to a fractional change in the concentration of the enzyme in question and its value ranges between 0 and 1.0. However, it is possible for an enzyme to have a flux control coefficient with a negative value as is the case at branch points where one metabolite has to be partitioned between two enzymes. In such a case the increase in flux through one branch is generally at the expense of the other.

The measurement of the flux control coefficient of a particular enzyme allows an accurate prediction of how flux through a given pathway might fluctuate in response to a specific change in its catalytic activity or its concentration. While a change in the concentration can be brought about by cloning and subsequent overexpression of the structural gene encoding the enzyme in question, changes in the catalytic activity of the enzyme without changing its concentration can be brought about through mutagenesis and protein engineering techniques. For example, consider pyruvate dehydrogenase (PDH) with the view of assessing its impact or influence on flux to acetate excretion in *E. coli* (Figure 7.1). The influence of PDH in that direction can be assessed from the enzyme's flux control coefficient, which can be calculated from the tangent to the curve of a log–log plot of flux (J) as a function of enzymic activity or concentration (E). Assuming that a small increase in the concentration of PDH

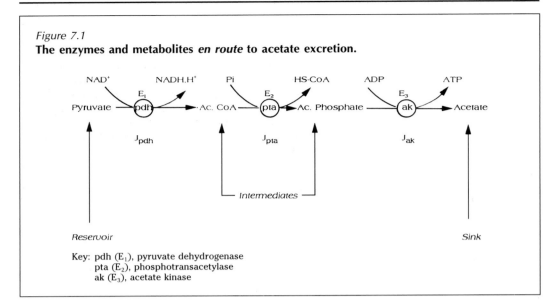

Figure 7.1

The enzymes and metabolites *en route* to acetate excretion.

Key: pdh (E$_1$), pyruvate dehydrogenase
pta (E$_2$), phosphotransacetylase
ak (E$_3$), acetate kinase

(dE_{pdh}) was accompanied by a small increase in the steady state flux (J) of the enzyme acetate kinase (dJ_{ak}), it follows, if we were to change the concentration of PDH very slightly, then the ratio dJ_{ak}/dE_{pdh} becomes equal to the slope of the tangent to the curve of J_{ak} against E_{pdh} as depicted in Figure 7.2. Analysing the data in this way, however, is somewhat imperfect as the numerical value of enzyme concentration and units of enzymic activity will be different from one enzyme to another. This problem could be overcome if we were to relate the fractional changes in flux through acetate kinase to the fractional increase in the concentration of PDH, i.e. dJ_{ak}/J_{ak} and dE_{pdh}/E_{pdh}, and as such the flux control coefficient will assume a value between 0 and 1.0 which can then be expressed in terms of a percentage.

However, it is possible for an enzyme to have a flux control coefficient with a negative value, as is the case at branch points where one metabolite has to be partitioned between two enzymes. In such a case the increase in flux through one branch is generally at the expense of the other, as exemplified in the case study for the partition of carbon flux at the junction of isocitrate (see section 7.3). At this junction, any increase in the concentration of isocitrate dehydrogenase (ICDH) is concomitant with a decrease in flux through the competing enzyme, namely isocitrate lyase (ICL). It is possible, therefore, to describe ICDH as having a negative flux control coefficient on flux through ICL. While any increase in the concentration of ICDH is accompanied by a decrease in flux through ICL, the opposite is not true for reasons which will become apparent later on; for further details see El-Mansi *et al.* (1994).

Figure 7.2

Determination of the flux control coefficient of pyruvate dehydrogenase (pdh) with respect to acetate excretion. The graph shows a typical pattern of variations in flux to acetate – measured as acetate kinase (J_{ak}) – in response to changes in the concentration of pyruvate dehydrogenase (E_{pdh}). Although $\log_{10}$ is used, natural logs can also be used providing that the data on the concentration of the enzyme and that on the flux are treated in the same way.

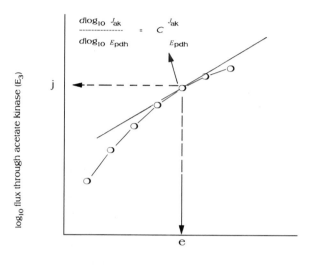

$$\frac{d\log_{10} J_{ak}}{d\log_{10} E_{pdh}} = C^{\,J_{ak}}_{\,E_{pdh}}$$

7.2.2 *The summation theorem*

The **summation theorem**, the second pillar of the MCA theory, states that the total sum of flux control coefficients of all enzymes in a given pathway adds up to 1.0. The summation theorem also shows that the flux control coefficient of an enzyme is a system property, because any increase in the concentration of a particular enzyme is accompanied by a decrease in its flux control coefficient. Such a decrease, according to the summation theorem, will have to be balanced by increasing the flux control coefficient of another enzyme – or more than one enzyme – within the same pathway so that the sum of all flux control coefficients remains constant, i.e. 1.0. For example, in a linear pathway consisting of enzymes with usual kinetic properties (i.e. where substrates stimulate and products inhibit reaction rate), the flux control coefficients for every enzyme must be 0 or higher, with a total sum of 1.0. If an enzyme were to show a flux control coefficient of 1.0 with all other enzymes showing flux control coefficients of 0, such an enzyme could justifiably be described as rate-limiting. The summation theorem also shows that this is not necessarily the case, because it is also possible for some or all of the enzymes to have values greater than 0 providing that the total does not exceed 1.0. In practice, we would expect a pathway flux to be influenced mainly by enzymes in that pathway, and to a much lesser extent by closely related pathways, and that distantly connected enzymes would have negligible influence or none at all. In other words, the flux control coefficients of hundreds or even thousands of enzymes which are not directly related or connected to the pathway in question will be zero despite the fact that flux control is shared among all enzymes.

Another consequence of the highly branched and intricate nature of cellular metabolism is that the central pathways provide biosynthetic precursors and energy for other pathways. So, as biosynthetic precursors are made, some are fed directly into the biosynthetic routes which in turn diminishes flux through the central metabolic pathways. It follows, therefore, that biosynthetic enzymes are likely to have negative flux control coefficients with respect to flux through the central metabolic pathways. According to the summation theorem if one or more enzymes possess a negative value of flux control coefficient, then it is possible to see some other enzymes displaying a flux control coefficient higher than the numerical value of 1.0. This is because if there are negative flux control coefficients, one or more flux control coefficients would have to be greater than 1.0 so that the total sum adds up to 1.0. This shows that the flux control coefficient is not an intrinsic property of the enzyme itself but rather a property of the whole system.

7.2.3 *Elasticities*

The flux control coefficient of an enzyme is influenced by the enzyme's ability to respond to changes in the concentration of its immediate substrate, as well as its ability to influence the concentrations of other metabolites in the pathway, a linkage which was first demonstrated by Heinrich and Rapaport (1974). The **elasticity coefficient**, the third pillar of the MCA theory, was therefore introduced to describe how flux is influenced by changes in the concentration of a given metabolite. In other words, **elasticity** is a parameter which describes, in quantitative terms, the sensitivity and responsiveness of an enzyme to a metabolite.

Unlike the flux control coefficient, elasticity is a property of individual enzymes and not of the pathway. The elasticity of an enzyme to a metabolite is defined by the slope of the curve of enzyme units (reaction rate) plotted as a function of metabolite concentrations, with the measurements taken at the metabolite concentration found *in vivo*. By analogy with the flux control coefficient (Figure 7.2), the value of the elasticity coefficient, which can be calculated from the slope, will depend upon the units used for the measurement of enzymic activities which may vary from one enzyme to another. This can be avoided, as described earlier for the flux control coefficient, by calculating the elasticity coefficient directly from a log–log plot of catalytic activity versus metabolite concentration to give the fractional change in enzymic activity as a function of the fractional change in the concentration of the substrate. As highlighted in the case study presented in section 7.3, elasticities have positive values for metabolites which stimulate enzymic activity (substrates, activators) and negative values for those which decrease reaction rate, such as products and inhibitors. Elasticity is, therefore, a parameter which describes, in quantitative terms, the sensitivity and responsiveness of an enzyme to a particular metabolite which could be a substrate, a product or an effector.

7.2.4 *The connectivity theorem*

This theorem, the fourth pillar of the MCA theory, addresses the question of how the flux control coefficient of a given enzyme can be related to its kinetic properties. Such an interrelationship is governed by the **connectivity theorem** which states that the sum of all connectivity values in a given pathway is zero. The connectivity value for any given enzyme can be calculated by multiplying its flux control coefficient by its elasticity with respect to the metabolite in question. Naturally, enzymes which are not affected by the metabolite in question will have an elasticity of zero and as such will make no contribution towards the final sum obtained. Further analysis of connectivity values has revealed that large elasticities are associated with small flux control coefficients, and vice versa. The mathematical equations relating the connectivity theorem to linear pathways, branch points and cycles have been described and dealt with extensively elsewhere (Fell, 1997).

7.2.5 *Response coefficients*

Induction and repression of enzyme synthesis in response to internal or external environmental stimuli are widely distributed in Nature and are very effective in 'turning on' and 'switching off' transcription. Covalent modification through reversible phosphorylation is another mechanism which regulates the activity of existing enzymes by rendering them active or inactive, as is the case for isocitrate dehydrogenase (ICDH) in *Escherichia coli* during adaptation to acetate (Koshland, 1987; Cozzone,

1988). In addition to degradation of mRNA and proteins, enzymes may also be the subject of allosteric control mechanisms which change the enzyme's affinity towards its substrate and/or co-factor(s).

The **response coefficient**, the fifth pillar of the MCA theory, reflects the effectiveness of a particular effector on flux through a given pathway and is dependent on two factors, namely, the flux control coefficient of the target enzyme and the strength of the effector, given by its elasticity coefficient. Clearly, for an effector to have a significant effect on flux, each of the above parameters with respect to the target enzyme has to be of a value higher than zero. Furthermore, in circumstances where more than one effector are involved, their response coefficients would all contain the same flux control coefficient as a component term. The flux control coefficient is, therefore, a valuable parameter as it indicates the relative strength and the capacity enjoyed by the enzyme for the control of metabolic flux.

Under circumstances where a particular effector may activate or inactivate more than one enzyme in a given pathway, the total response will be the sum of the individual responses from each enzyme affected (Hofmeyr and Cornish-Bowden, 1991). However, this is only true when the changes in the concentration of the effector are very small because of the non-linear relationship of the kinetics in metabolic systems.

Now, let us consider how carbon flux is partitioned at the junction of isocitrate during growth of *Escherichia coli* on acetate.

7.3 Effect of isocitrate dehydrogenase (ICDH) and isocitrate lyase (ICL) on the partition of carbon flux at the junction of isocitrate during growth of *Escherichia coli* on acetate: a case study

During growth on acetate as sole source of carbon and energy, *E. coli* requires the operation of the anaplerotic sequence of the glyoxylate bypass for the provision of biosynthetic precursors (Kornberg, 1966). Under these conditions a new junction is generated at the level of isocitrate (Figure 7.3) where isocitrate lyase (ICL) of the glyoxylate bypass is in direct competition with the Krebs cycle enzyme isocitrate dehydrogenase (ICDH). Although ICDH has a much higher affinity for isocitrate, flux through ICL and thence the anaplerotic enzyme malate synthase (MS) is assured by virtue of high intracellular levels of isocitrate and the

Figure 7.3

A diagrammatic representation of the central metabolic pathways employed by *Escherichia coli* for the metabolism of acetate, highlighting the direct competition between isocitrate lyase (ICL) of the glyoxylate bypass and the Krebs cycle enzyme isocitrate dehydrogenase (ICDH) for their common substrate. It also highlights the bifunctional role of isocitrate dehydrogenase kinase/phosphatase in the moiety-conserved cycle involved in the reversible inactivation of ICDH.

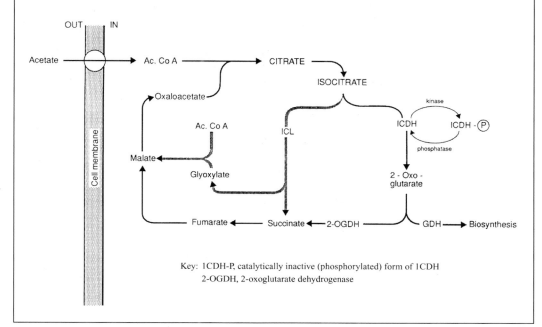

Key: ICDH-P, catalytically inactive (phosphorylated) form of ICDH
2-OGDH, 2-oxoglutarate dehydrogenase

partial inactivation of ICDH (El-Mansi *et al.*, 1985). Although the *in vivo* signal which triggers the expression of the glyoxylate bypass is yet to be determined, recent investigations have revealed that acetate *per se* can be safely ruled out as a possible signal (El-Mansi, 1998). Using radiolabelled isotopes and NMR spectroscopy, Walsh and Koshland (1984) have been able to quantitatively determine the rate of carbon flux through the two branches at this junction.

In order to assess the relative contribution of each of the above enzymes to the overall distribution of carbon flux among various enzymes of central metabolism, the computer software package MetaModel was used to calculate the steady-state fluxes and the concentration of various metabolites during growth of *Escherichia coli* on acetate. This computer package also enabled us to formulate the matrices of the elasticity coefficients and the control and response coefficients under different steady states. In the next section, we will discuss the data in the light of Kacser's MCA theory as well as the traditional concept of the rate-limiting step.

7.3.1 *The model*

The complex enzyme system used by Walsh and Koshland (1984) for the central pathways was reduced to a skeleton model (Figure 7.4) in order to explore the consequences of controlled adjustment of ICL and ICDH enzymic activities on the partition of carbon flux among various enzymes of central metabolism. MetaModel 2.1, written and compiled by Cornish-Bowden and Hofmeyr (1991) was used to solve the steady state and to calculate the various coefficients. This computer program enabled the simulation of a whole series of variables and any small change in any variable was detected, transmitted systematically and all the changing fluxes, metabolites and coefficients were calculated accordingly.

From this simple multi-enzyme system (Figure 7.4), differential equations were derived to describe the rate of change in the concentration of each metabolite as it is converted by the

---

Figure 7.4

A skeleton model describing the central metabolic pathways of *Escherichia coli* during growth on acetate. Metabolites and enzymes are as described below.

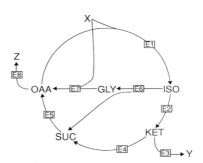

Metabolites:
X   = acetyl-CoA   (reservoir excluding flux to fatty acids)
ISO = isocitrate
KET = a-ketoglutarate
SUC = succinate
OAA = oxaloacetate
GLY = glyoxylate
Y   = glutamate   (sink for cell constituents)
Z   = gluconeogenesis and biosynthesis   (sink for cell constituents)

Enzymes:
E1 = citrate synthase/aconitase
E2 = isocitrate dehydrogenase
E3 = glutamate dehydrogenase
E4 = a-ketoglutarate dehydrogenase/succinate thiokinase
E5 = succinate dehydrogenase/fumarase/malate dehydrogenase
E6 = isocitrate lyase
E7 = malate synthase
E8 = PEP carboxykinase

Table 7.1 Balance equations for the skeleton model proposed for the central metabolic pathways of *E. coli* during growth on acetate

Balance equation	Flux relationships	
0 = dISO/dt = V1 – V2 – V6	> V1 = V2 + V6	
0 = dKET/dt = V2 – V4 – V3	> V2 = V3 + V4	> V3 = V2 – V4
0 = dSUC/dt = V4 + V6 – V5	> V5 = V4 + V6	> V4 = V5 – V6
0 = dGLY/dt = V6 – V7	> V6 = V7	
0 = dOAA/dt = V5 + V7 – V8 – V1	> V1 + V8 = V5 + V7	> V8 = V5 – V2

enzymes (Table 7.1). In a steady state, the rate of change for each metabolite equals zero, i.e. the net rate of formation equals the rate of consumption. It is clear from the skeleton model that we have eight different reactions, of which the velocities of conversion through enzymes 6, isocitrate lyase, and 7, malate synthase, are identical due to their linear relationship.

It is noteworthy that the model depicted above for the Krebs cycle and the glyoxylate bypass (Figure 7.4) does not constitute a moiety conserved cycle (Hofmeyr *et al.*, 1986) because there are reversible sinks to the cycle as well as branching from within the cycles, thus giving rise to metabolite pools which are involved in three different reactions.

In formulating the reaction equations in our model, the following reversible Michaelis–Menten type equation was used as a basis to describe the enzymic reactions:

$$v = \frac{V_f{}^*S/K_s - V_r{}^*P/K_p}{1 + S/K_s + P/K_p}$$

where V_f and V_r are the V_{max} values of the forward and the reverse reactions respectively; S = substrate; P = product, and K_s and K_p are the Michaelis–Menten constants, i.e. K_m values for the relevant metabolites.

The differential equations which describe the rate of change of each substrate concentration in the central pathways following the entry of acetyl-CoA are shown in Table 7.2. In this model we have taken into account the differing affinities of ICDH and ICL for their common substrate. Walsh and Koshland (1984) established that reversible inactivation of ICDH during growth on acetate allowed one-third of the flux through ICL (i.e. a ratio of 2.6 : 1 in favour of ICDH). The concentration of E6 (ICL) in our model was, therefore, set at 0.388 whilst all other enzyme concentrations remained at 1.0. Furthermore, the intracellular concentrations of acetyl-CoA, glutamate and phosphoenolpyruvate (PEP) as 'external' metabolites were fixed at concentrations of 56, 1.06 and 1.69 mM, respectively. These values represent the input of acetyl-CoA and the outputs of glutamate

Table 7.2   Reaction expressions and rate equations for the skeleton model proposed for the central pathways of *E. coli* during growth on acetate

---

1.   X + OAA = ISO
     V/[E1] = (5.1*X*OAA – ISO)/(1 + X + OAA + ISO)

2.   ISO = KET
     V/[E2] = (10*ISO – KET)/(1 + ISO + KET)

3.   KET = Y
     V/[E3] = (7.5*KET – Y)/(1 + KET + Y)

4.   KET = SUC
     V/[E4] = (10*KET – SUC)/(1 + KET + SUC)

5.   SUC = OAA
     V/[E5] = (5*SUC – OAA)/(1 + SUC + OAA)

6.   ISO = GLY + SUC
     V/[E6] = (10.6*ISO – SUC*GLY)/(1 + ISO + SUC + GLY)

7.   GLY + X = OAA
     V/[E7] = (10*X*GLY – OAA)/(1 + X + GLY + OAA)

8.   OAA = Z
     V/[E8] = (11*0AA – Z)/(1 + OAA + Z)

---

and oxaloacetate required for the biosynthesis of one gram dry weight biomass of *E. coli*, as described previously (El-Mansi *et al.*, 1994). The steady-state growth of *E. coli* on acetate was simulated and the data obtained (Table 7.3) are in good agreement with the *in vivo* data reported by Walsh and Koshland (1984).

The changes in the steady-state fluxes through ICDH and ICL as well as the intracellular concentration of isocitrate in response to changes in the concentration of ICDH (E2) are shown in Figure 7.5. As the rate of carbon flux through ICDH is diminished, the intracellular concentration of isocitrate rises to a level

Table 7.3   Steady-state fluxes and pools as calculated by MetaModel for the skeleton model (Figure 7.4) representing the central metabolic pathways of *E. coli* during growth on acetate

Enzyme	Velocity	Metabolites (mM)	Velocity
[E1] = 1.0000	1.1117	[x] = 56.0000	fixed
[E2] = 1.0000	0.8008	[OAA] = 0.2236	variable
[E3] = 1.0000	0.0468	[ISO] = 0.1174	variable
[E4] = 1.0000	0.7540	[KET] = 0.1551	variable
[E5] = 1.0000	1.0469	[Y] = 1.0600	fixed
[E6] = 0.3880	0.3109	[SUC] = 0.3880	variable
[E7] = 1.0000	0.3109	[GLY] = 0.0322	variable
[E8] = 1.0000	0.2642	[Z] = 1.6900	fixed

All velocities are expressed in terms of mmol of substrate consumed / min / 10 ml of cell volume.

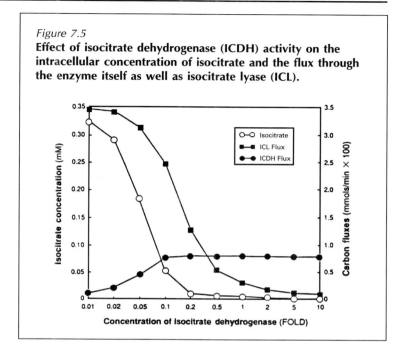

Figure 7.5
Effect of isocitrate dehydrogenase (ICDH) activity on the intracellular concentration of isocitrate and the flux through the enzyme itself as well as isocitrate lyase (ICL).

that sustains flux through ICL despite its low affinity. The data also show that any decrease in the concentration of ICDH results in an increase of carbon flux through ICL and, in turn, the anaplerotic enzyme malate synthase (MS). This in turn replenishes the central pathways with primary intermediates and biosynthetic precursors. During steady-state growth on acetate, the rate of carbon flux through ICL is 31 mmoles per min; this represents 33% of the total carbon processed at this junction.

According to the summation theorem, the flux through ICDH can be affected by other enzymes of the central pathways and the sum of the flux control coefficients of all enzymes on J_{ICDH} is 1.0. While the flux control coefficient is a measure of the relative change in flux through a particular enzyme in a given pathway, in response to a small change in its concentration, the elasticity coefficient on the other hand is a measure of the relative change in flux in response to a small change in the concentration of the substrate or a given co-factor. Tables 7.4 and 7.5 show the matrices of the flux control coefficients and the elasticity coefficients as calculated by MetaModel respectively for the enzymes of the central metabolic pathways during steady-state growth on acetate. The data outlined in Figure 7.6 clearly show that any increase in the concentration of ICDH above the level observed in acetate phenotype, i.e. 1.0, does not increase flux through the enzyme itself nor does it increase the enzyme's flux control coefficient. On the other hand, however, any drop in the

Table 7.4   Flux control coefficients as calculated by MetaModel for the central metabolic pathways of *E. coli* during steady-state growth on acetate

Enzyme	J1	J2	J3	J4	J5	J6/J7	J8
E1	0.8599	0.8944	7.5300	0.4829	0.5670	0.7708	−0.4256
E2	−0.2399	−0.0164	−1.1192	0.0520	−0.2013	−0.8156	−0.7619
E3	−0.0257	−0.0265	0.2985	−0.0466	−0.0400	−0.0239	−0.0809
E4	0.1797	0.2359	−2.9514	0.4336	0.3172	0.0349	0.5635
E5	0.1996	0.1617	−1.5563	0.2682	0.2767	0.2971	0.6251
E6	0.2760	0.0224	1.3018	−0.0570	0.2309	0.9291	0.8631
E7	0.0083	0.0007	0.0400	−0.0017	0.0070	0.0285	0.0265
E8	−0.2579	−0.2722	−2.5432	−0.1314	−0.1575	−0.2208	0.1902
Sums	1.0000	1.0000	1.0000	1.0000	1.0000	1.0000	1.0000

Table 7.5   Elasticity matrices as calculated by MetaModel for the central metabolic pathways of *E. coli* during steady-state growth on acetate

Enzyme	X	OAA	ISO	KET	Y	SUC	GLY	Z
E1	0.0252	0.9979	-3.9×10^{-3}	0	0	0	0	0
E2	0	0	1.0600	−0.2742	0	0	0	0
E3	0	0	0	11.1642	−10.7127	0	0	0
E4	0	0	0	1.2329	0	−0.5849	0	0
E5	0	−0.2690	0	0	0	−0.8896	0	0
E6	0	0	0.9388	0	0	−0.2625	−0.0311	0
E7	0.0345	−0.0165	0	0	0	0	1.0120	0
E8	0	3.1190	0	0	0	0	0	−2.7707

intracellular level of ICDH activity beyond that of acetate phenotype (1.0) appears to have a profound effect on carbon flux through the enzyme itself and the enzyme's flux control coefficient. Clearly, the reduction of ICDH activity to 20% or less (Figure 7.4) led to a sharp increase in the flux control coefficient of ICDH and this was mirrored in the rate of carbon flux through the enzyme itself. It follows therefore that flux through ICDH during growth on acetate is in excess of cellular demands and as such cannot be rate limiting.

Modulation of ICL activity directly by systematically increasing the concentration of the enzyme revealed some interesting observations. From this simulation (Figure 7.7) the indications were that above a certain threshold of ICL concentration the two cycles work in concert and the partition of carbon flux between ICDH and ICL is no longer a problem. It is interesting that increasing the concentration of ICL resulted in an increase of flux through ICDH as well as ICL and that this was inevitably

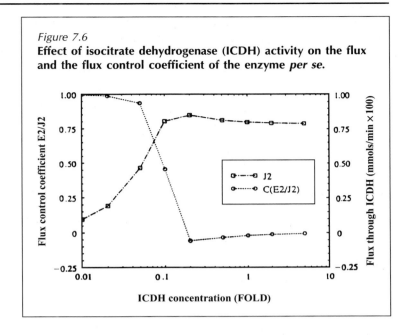

Figure 7.6
Effect of isocitrate dehydrogenase (ICDH) activity on the flux and the flux control coefficient of the enzyme *per se*.

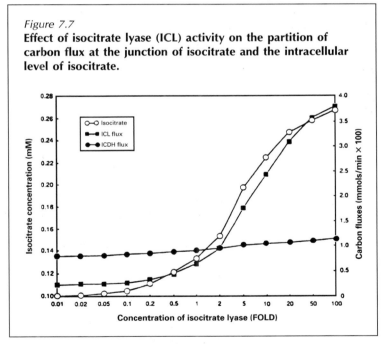

Figure 7.7
Effect of isocitrate lyase (ICL) activity on the partition of carbon flux at the junction of isocitrate and the intracellular level of isocitrate.

at the expense of the intracellular concentration of isocitrate which only rose to 0.271 mM. At an ICL concentration of 100, the rate of carbon flux through ICDH rose to 115 mmols/min, while that through ICL rose to 378 mmols/min. From the above data, we have been able to assess how perturbation of a given enzymic activity is transmitted throughout the various components of central metabolism. The data demonstrate that increasing the concentration of ICDH beyond the level observed in the wild type was not accompanied by an increase in flux through ICDH *per se* or the central pathways as a whole. Furthermore, increasing the concentration of ICDH results in depletion of the intracellular pool of isocitrate, thus diminishing flux through the glyoxylate bypass.

Summary

- The qualitative terms used to describe an enzyme as rate-limiting, bottleneck or pacemaker should be abandoned and replaced by the term **rate controlling** and further qualified by the **flux control coefficient** which describes, in quantitative terms, the capacity of the enzyme with respect to flux control.

- The flux control coefficient of a particular enzyme in a given pathway is a **system property**, i.e. its value is not entirely independent of other enzymes in the pathway. The interrelationship between the flux control coefficient of a particular enzyme and that of other enzymes in a given pathway is governed by the **summation theorem** which dictates that the total sum of all flux control coefficients in a given pathway adds up to 1.0.

- In a steady state, the influence of a particular metabolite on flux through a given enzyme on the one hand and the whole pathway on the other can be determined from the enzyme's **elasticity coefficient**, a quantitative term that is directly related to the kinetic properties of the enzyme involved.

- The metabolic interrelationship between the flux control coefficient and the elasticity coefficient is described by the **connectivity theorem** which takes into account the kinetic properties of each of the enzymes involved.

- The action of external **effectors** on metabolic flux can be assessed by measuring the **response coefficient** which, to a large extent, is dependent on the flux control coefficient and the elasticity coefficient of the enzyme with respect to the effector. For an *effector* to be able to influence the flux through a certain enzyme, the values of the aforementioned coefficients must be relatively high.

- Activating or increasing the catalytic activity of a single enzyme is not usually accompanied by a significant increase in flux (productivity) even with enzymes possessing a relatively large flux control coefficient. This is simply because the flux control shifts to other enzymes as the target enzyme is activated. This, in turn, implies that amplification of single enzymic activity is not a viable option for biotechnologists who wish to increase the productivity of a given pathway. This limitation, however, does not apply to the reduction of catalytic activity as reduction or inactivation is generally accompanied by a considerable drop in flux.

- Increasing the concentration of an effector which activates all enzymes in the pathway will be accompanied by an appreciable increase in flux and, in turn, yield. Furthermore, an effector which stimulates the activity of more than one enzyme in a given pathway may lead to increase in flux (productivity) particularly if those enzymes share a relatively high flux control coefficient.

- The case study presented in this chapter for ICDH and ICL revealed that during growth on acetate as sole source of carbon and energy, isocitrate dehydrogenase (ICDH) is not rate controlling and that flux through isocitrate lyase (ICL) is essential not only to replenish the central pathways with biosynthetic precursors but also to sustain a high intracellular level of isocitrate. Furthermore, metabolic simulation revealed that above a certain threshold concentration of ICL, the Krebs cycle and the glyoxylate bypass work in concert.

References

Cornish-Bowden, A. and Hofmeyr, J.H.S. (1991) MetaModel: a program for modelling and control analysis of metabolic pathways on the IBM PC and compatibles. *Comp. Appl. Biosci.*, **7**, 89–93.

Cozzone, A.J. (1988) Protein phosphorylation in prokaryotes. *Ann. Rev. Microbiol.*, **42**, 97–125.

El-Mansi, E.M.T. (1998) Control of metabolic interconversion of isocitrate dehydrogenase between the catalytically active and inactive forms in *Escherichia coli, FEMS Microbiol. Lett.*, **166**(2), 333–339.

El-Mansi, E.M.T. and Holms, W.H. (1989) Control of carbon flux to acetate excretion during growth of *Escherichia coli* in batch and continuous cultures. *J. Gen. Microbiol.*, **135**, 2875–2883.

El-Mansi, E.M.T., Dawson, G.C. and Bryce, C.F.A. (1994) Steady-state modelling of metabolic flux between the tricarboxylic acid cycle and the glyoxylate bypass in *Escherichia coli, Comp. Appl. Biosci.*, **10**, 295–299.

El-Mansi, E.M.T., Nimmo, H.G. and Holms, W.H. (1985) The role of isocitrate in control of the phosphorylation of isocitrate dehydrogenase in *Escherichia coli, FEBS Lett.*, **183**, 251–255.

Fell, D. (1997) *Understanding the Control of Metabolism*, London: Portland Press.

Gubler, M., Park, S.M., Jetten, M., Stephanopoulos, G. and Sinskey, A.J. (1994) Effect of phosphoenolpyruvate carboxylase deficiency on metabolism and lysine production in

Corynebacterium glutamicum, Appl. Microbiol. Biotechnol., 40, 857–863.

Heinrich, R. and Rapaport, T. (1974) A linear steady-state treatment of enzymatic chains. *Eur. J. Biochem.*, 42, 89–95.

Hofmeyr, J.H.S. and Cornish-Bowden, A. (1991) Quantitative assessment of regulation in metabolic systems. *Eur. J. Biochem.*, 200, 223–236.

Hofmeyr, J.-H.S., Kacser, H. and Merwe, K.J. (1986) Metabolic control analysis of moiety-conserved cycles. *Eur. J. Biochem.*, 155, 631–641.

Holms, W.H. (1996) Flux analysis and control of the central metabolic pathways in *Escherichia coli, FEMS Microbiol. Rev.*, 19, 85–116.

Kacser, H. and Burns, J. (1973) The control of flux, *Symp. Soc. Exp. Biol.*, 27, 65–104. (reprinted in *Biochem. Soc. Trans.*, 23, 341–366, 1995)

Kornberg, H.L. (1966) The role and control of the glyoxylate cycle in *Escherichia coli, Biochem. J.*, 99, 1–11.

Koshland, D.E. Jr (1987) Switches, thresholds and ultrasensitivity. *Trends Biochem. Sci.*, 12, 225–229.

Savageau, M.A., Voit, E.O. and Irvine, D.H. (1987) Biochemical systems theory and metabolic control theory: 1, Fundamental; similarities and differences. *Math. Biosci.*, 86, 127–145.

Stephanopoulos, G. and Vallino, J.J. (1991) Network rigidity and metabolic engineering in metabolite overproduction. *Science*, 252, 1675–1681.

Vallino, J.J. and Stephanopoulos, G. (1993) Metabolic flux distributions in *Corynebacterium glutamicum* during growth and lysine overproduction. *Biotechnol. Bioeng.*, 41, 633–646.

Walsh, K. and Koshland, D.E. (1984) Determination of flux through the branch point of two metabolic cycles: the tricarboxylic acid cycle and the glyoxylate shunt, *J. Biol. Chem.*, 259, 9646–9654.

8 Biosensors in Bioprocess Monitoring and Control

Marco F. Cardosi

8.1 Introduction

Fermentation processes by their very nature are complex systems, especially for the purpose of effective bioprocess monitoring. Typically, the biological system (e.g. cell or enzyme) is surrounded by a chemical and physical environment where changes in one of these environments can cause drastic effects on the other parts of the system as a whole. The ultimate aim of bioprocess analysis therefore is a detailed monitoring of the biological system, the chemical and physical environment and how these interact.

The current state of sensor technology permits the measurement and control of dissolved oxygen, dissolved carbon dioxide, pH, redox potential, temperature, agitation and the level of foam in the fermentor vessel. Recent advances in sensor technologies have led to the on-line determination of biomass through *in situ* optical density or fluorometric probes and capacitance measurements using low radio frequencies (Biomass monitor, Aber Instruments Ltd, Aberystwyth, UK). Oxygen uptake and/or carbon dioxide evolution during cell growth have been monitored using piezoelectric mass balances and off-gas analysis. In addition, head space analysis of volatiles such as methanol has been used to monitor and control fermentative processes.

However, no reliable techniques exist to carry out real-time measurements of non-volatile substrates and metabolites in the reactor vessel. In this respect, biosensors have attracted a considerable amount of attention over the last decade because they offer the possibility to monitor single analytes even in the extremely complex and often undefined matrices that occur in fermentation vessels; however, the commercial exploitation of biosensors for fermentation process monitoring and control has not yet been fully realized.

8.2 Basic components of on-line process monitoring and control

Any system for on-line monitoring and control must include three essential elements: (1) a sensor (in the context of this chapter this is a biosensor), (2) a suitable analysis manifold employing an *in situ* or *ex situ* arrangement for contacting the fermentation liquor with the bioprobe and (3) a control system with the necessary hardware and control algorithms in order to employ a suitable control strategy. In order to use the analysis data efficiently, the time lag between sampling and analysis must be within the timescales of the bioprocess. Thus, in a bacterial cultivation process with short generation times, the time delay should be in the order of a very few minutes. In cultivation processes employing much slower growing mammalian cells, this time delay can be in the order of 1–2 hours. In a reactor employing an immobilized enzyme, the time delay should be of the order of seconds. The time delay is a function of the sampling, the sample handling, the analysis and the data process.

Sensors can be interfaced to a biotechnological process in different ways. A biosensor can be used as an *in situ* probe, or can be separated from the fermentation broth by a filtration unit, e.g. tangential flow units in a bypass. An alternative and perhaps more practical way of utilizing biosensors in continuous fermentation monitoring and control is to use them in conjunction with flow injection analysis (FIA). In FIA, the liquid sample to be analysed, which forms a discrete zone, is injected into a moving, non-segmented carrier stream flowing continuously past a detector (Ruzicka and Hansen, 1988). The many advantages of this technique when applied to process control include:

- reduced risk of contamination because the sample is not returned to the bioreactor
- automatic recalibration of the sensor to counter any problems associated with drift
- short response time
- requirement for low sample volumes
- multi-component monitoring if an array of biosensors is used as the detector.

Certain practical considerations need to be taken into account when using biosensors for process monitoring. With the use of *in situ* biosensors, sterilization of the probe can be a major problem that is usually solved by a complex engineering design. Furthermore, some biocatalysts may require very strict operational conditions that may be different from those in the fermentor. With FIA, the associated problems are different and include: faults associated with in-line membranes used to produce the cell-free samples for analysis, microbial growth on the upstream side of the membrane causing plugging, need for dilution of the sample or adjustment of pH prior to the analysis step, dissolved gas in the sample or flow stream and drift or deactivation of the sensor with time.

Box 8.1 **Biosensors**

Biosensors belong to a subgroup of chemical sensors in which a biologically based mechanism is used for analyte detection. The International Union of Pure and Applied Chemistry (IUPAC) has defined chemical sensors as miniaturized transducers that reversibly and selectively respond to chemical compounds or ions yielding an analytical signal proportional to the analyte concentration. By definition, a biosensor is an analytical device that combines the specificity of a biological sensing element for the analyte of interest with a transducer to produce a signal proportional to target analyte concentration.

This signal can result from: a change in pH, release or uptake of gases, light emission, heat emission, mass change, etc. The transducer converts the biological reaction into a measurable response such as current, potential, temperature change, modulation of light intensity, etc. The signal can be further amplified, processed or stored for later analysis and retrieval. Table 8.1 lists the different components that can be used to construct a biosensor. In principle, any receptor can be combined with any suitable transducer to produce a working probe. Figure 8.1 shows a generalized representation of a biosensor.

Table 8.1 Components that can be used in the construction of a biosensor

Biological element	Transducer
Whole organisms	Potentiometric
Tissues	Amperometric
Cells	Conductometric
Organelles	Impedimetric
Membranes	Optical
Enzymes	Calorimetric
Receptors	Acoustic
Antibodies	Mechanical
Genetic material	
Organic molecules	

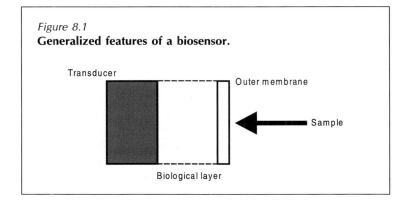

Figure 8.1
Generalized features of a biosensor.

In terms of process monitoring and control applications, the emphasis has been on the use of enzymes, and in particular in an enzyme electrode configuration. (The enzyme electrode is a combination of any type of electrochemical probe with a thin layer (10–200 μm) of immobilized enzyme.) Typically, the progress of a particular enzymic reaction (which is related to the concentration of analyte) is measured by monitoring the rate of product formation or the rate of disappearance of a reactant. If either the product or reactant is electroactive, the reaction may be monitored directly using amperometry. Consequently, this trend will be reflected in the subject matter of this chapter although other successful strategies will also be discussed.

8.3   Enzymes

Enzymes are very selective biological catalysts that carry out the conversion of a particular substrate into a product under mild operating conditions, i.e. room temperature, low ionic strength and near neutral pH. An enzyme will usually catalyse a single chemical reaction or a set of closely related chemical reactions. Side reactions leading to wasteful formation of by-products rarely occur. Thus, quantitative assays may be done on crude materials with little or no sample preparation.

Much of the catalytic power of enzymes comes from their ability to bring substrates together in favourable geometric alignments to form enzyme–substrate complexes. The substrates are bound to a specific region of the enzyme known as the active site. Most enzymes are highly specific in the binding of their substrates and indeed, the catalytic specificity of an enzyme is dependent at least in part on the specificity of this binding.

For many enzymes, the rate of catalysis (v), defined as the number of moles of product formed per second, varies with substrate concentration [S] as illustrated in Figure 8.2. At a fixed concentration of enzyme, v is directly proportional to [S] when the concentration is small. At high substrate concentrations, the rate of reaction becomes independent of [S] and reaches a maximum value. The model proposed, which is the simplest one to account for these kinetic properties is shown below:

$$E + S \underset{k_{-1}}{\overset{k_1}{\rightleftarrows}} ES \overset{k_2}{\longrightarrow} E + P$$

According to this model, an enzyme molecule E combines with substrate to form an ES complex, with a rate constant k_1. The ES complex then has two possible fates; it can dissociate back to E and S, with a rate constant k_{-1} or it can proceed in an irreversible manner to form a product P, with a rate constant k_2. Using

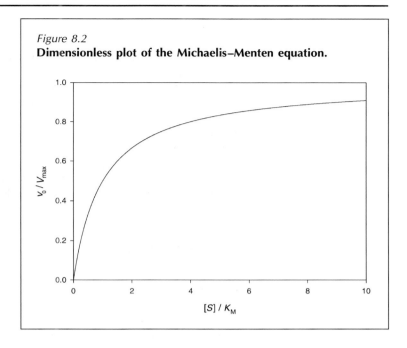

Figure 8.2
Dimensionless plot of the Michaelis–Menten equation.

this model, it is possible to define a rate equation (Michaelis–Menten equation) for an enzyme catalysed reaction that states:

$$v = v_{max}\frac{[S]}{[S] + K_M}$$

where K_M (the Michaelis constant) is defined as $\dfrac{k_{-1} + k_2}{k_1}$ and v_{max} is the product of the total enzyme concentration and k_3. Thus, when $[S] \ll K_M$ the rate of reaction v is defined as:

$$v = \frac{v_{max}[S]}{K_M}$$

i.e. the rate is directly proportional to $[S]$.

When using enzymes in biosensor applications, two important considerations have to be taken into account, i.e. operational stability and long-term use. Because both of these factors are to some degree a function of the immobilization strategy used, correct choice of immobilization technique is an important criterion. Many methods have been reported in the literature for the immobilization of enzymes to the surface of a transducer. These can be summarized as follows:

- Physical methods
  Entrapment in a gel matrix or polymer
  Adsorption to the surface of the transducer

- Chemical methods
  Covalent binding to the transducer surface using bifunctional reagents such as cyanuric chloride
  Crosslinking using multifunctional reagents such as glutaraldehyde
  Immobilization in an electrochemically grown polymer film

8.3.1 *Preparation of the immobilized enzyme layer*

The proper functioning of an enzyme-based sensor is, of course, heavily dependent on both the chemical and physical properties of the immobilized enzyme layer. To this end, there are many possible methods for immobilizing an enzyme at the surface of an electrode. For convenience, these can be divided into physical methods and chemical methods. An important consideration when developing an immobilization procedure is that the process should be applicable to a range of surfaces. This allows the choice of support to be as wide as possible and ensures that no refabrication of the support is required. Other advantages sought for in the immobilization method include: (a) an ability of the biological component to operate at a wider pH range than in solution; (b) attainment of greater stability resulting from the immobilization and (c) the generation of a defined 'diffusion region' on the surface of the electrode.

8.3.1.1 *Gel entrapment*

Entrapment within a 3-D gel matrix is a common method of enzyme immobilization. Numerous matrices have been employed but the most favoured have been alginate, crosslinked with linear chains of Ca^{2+} ions, gelatin and polyacrylamide. With due attention to the degree of crosslinking and the nature of the gelling process, i.e. minimizing the concentration of free radicals, gel entrapment can be applied to any enzyme. Unfortunately, this process does suffer from three major drawbacks: (1) large diffusional barriers to the transport of substrate or product resulting in reaction retardation and long response times, (2) continuous loss of enzyme activity because these materials generally do not have a narrow pore size distribution, and (3) shrinkage and/or swelling of the polymer depending upon the ionic strength of the milieu. A more straightforward method of enzyme entrapment is to retain the protein at the electrode surface behind a thin semi-permeable membrane such as dialysis tubing. Here, a thick paste of the enzyme in a suitable volume of buffer (1 or 2 mm^3) is spread over the surface of the electrode. The layer is then covered with a 20–25 μm thick dialysis membrane, of about 10 000 Daltons molecular mass cut-off, held in place by a suitably sized O-ring.

8.3.1.2 *Adsorption*

The great advantage of using adsorption as the immobilization technique is that usually no additional reagents are required and only a minimum of activation or 'clean up' need be done. Unfortunately however, only weak short-range interactions are involved such as Van der Waal forces, dipole–dipole interactions and hydrogen bonding. Because of this, the reversible nature of the binding equilibrium is highlighted by its susceptibility to changes in ambient conditions, i.e. pH, temperature, ionic strength, polarity, etc.

When a protein adsorbs to the surface of a support, the reaction will be influenced by non-covalent interactions between the surface of the protein and the support. Clearly the protein could interact with the surface in several different ways depending on the orientation with which it approaches the surface. In an unperturbed solution this process, and the reverse process, is under mass transport control. If every molecule that encounters the surface is adsorbed, a concentration gradient rapidly develops at the surface. The rate of adsorption then becomes proportional to the rate of diffusion such that:

$$\frac{dn}{dt} = C_0 \sqrt{\frac{D}{\pi t}}$$

where n is the number of molecules, C_0 the bulk concentration of protein and D the diffusion coefficient. The model does not however account for the situation where the layer next to the surface is only partially saturated with protein (as happens during sensor use). Under this condition, desorption becomes the dominant process and the biomolecule becomes lost from the electrode surface. Finally, Alvarez-Icaza and Schmid (1994) have shown that the adsorption of glucose oxidase to the surface of a carbon electrode leads to the irreversible denaturation of the protein unless steps are taken to modify the interface. Indeed, recent results in this laboratory (unpublished) have shown that the adsorption of the enzymes glucose oxidase and horseradish peroxidase to carbon does not proceed via a simple Langmuir isotherm. Rather, three equilibrium steps are involved suggesting that the protein exists in three different microcrystalline states on the surface thereby making the immobilized protein difficult to quantify. Despite these apparent shortcomings, the ease of the technique and its general applicability make adsorption particularly attractive as a pilot method or for the production of devices not requiring long-term stability, e.g. one-shot disposable sensors.

8.3.1.3 *Covalent immobilization*

Covalent binding of the enzyme to the surface of the electrode is generally irreversible and as such is the most stable method.

Chemical bonding to the surface must be effected by using nucleophilic amino acid (carboxylic acid, hydroxy, thiol, imidazole and phenolic groups) of the protein that are not involved in the biological function of the molecule. In general terms, the attachment of an enzyme to an electrode surface (e.g. a metal or carbon disc) is a two-step process. The first step involves activating the surface of the electrode, i.e. imparting some useful chemical reactivity to the otherwise inert surface. The second step involves the binding of the enzyme to the chemically activated electrode surface.

Immobilization of enzymes to metal electrodes
Although metal surfaces are in themselves unreactive, when they are covered by a thin surface oxide film they can be functionalized by reagents such as chloro or alkyl silanes (Murray, 1980). By analogy with silica surfaces, Pt/PtO, Au/AuO and SnO_2 surfaces have many M–OH sites (where M is the metal) and when they are contacted with, for example, a solution of dichlorodimethylsilane $[Cl_2Si(CH_3)_2]$ under anhydrous conditions the organosilane reagent becomes immobilized by formation of chemically stable –MOSi– bonds.

The silanization reaction can be monitored using surface analysis techniques such as X-ray photoelectron spectroscopy (XPS). Here, the experimenter would detect the silanization reaction of a 'clean' metal surface by noting the appearance of new XPS Si 2p, Si 2s and O 1s bands in the spectrum. Alternatively, other techniques such as reflectance FTIR may also prove useful if the modifying agent contains a suitable chromophore. Finally, it may also be possible to monitor the surface modification by measuring changes in the double layer capacitance due to the replacement of polar surface groups by non-polar methyl groups. Accurate measurements of the change in the double layer capacitance may allow quantization of the surface groupings. Although the reaction shown above produces a functionalized electrode surface, the resultant electrode is not suitable for further synthetic modification due to the chemical inertness of the methyl groups. To bond an enzyme to a silanized surface, it is important that the organosilane reagent itself bears chemical functionality such as a primary amine group or a carboxylic acid. A particularly useful reagent in this context is propyl amino silane $[(CH_3CH_2O)_3Si(CH_2)_3NH_2]$ which will not only functionalize a metal electrode but will then allow coupling chemistry to take place with proteins through the attendant amine grouping. A typical synthetic scheme using this reagent is shown in Figure 8.3, by which an enzyme is coupled to a silanized metal electrode via the formation of an amide bond between the amino group of the silane and a surface carboxylic acid on the protein. Note however, that to make the amide linkage, the carboxylic

Figure 8.3
Two-step synthetic scheme for the attachment of a protein to a metal electrode. In step (a), the metal oxide surface is functionalized with an aminosilane reagent. In step (b), a carbodiimide-activated protein is attached to the electrode surface via the attendant primary amine groups.

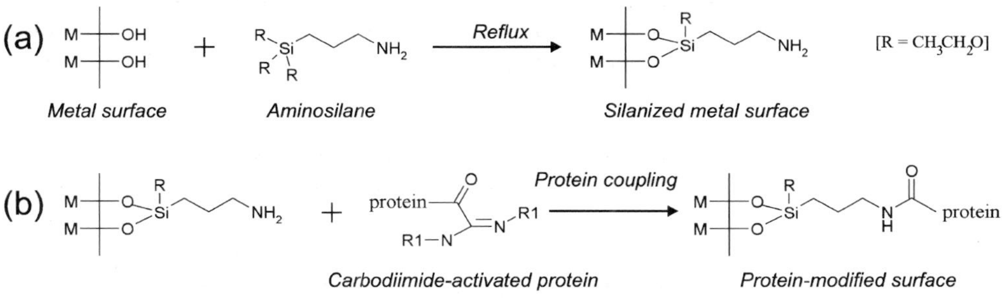

acid on the protein must itself be activated. This is normally achieved by using carbodiimide chemistry. The reaction uses attractively mild conditions and can draw upon a variety of diimide reagents developed largely for solid phase peptide synthesis. The incorporation of the enzyme onto the electrode surface can be monitored by challenging the electrode with the substrate of the enzyme and detecting the production of an electrochemical product.

Immobilization of enzymes to carbon electrodes
Carbon electrodes represent another popular material for the manufacture of enzyme electrodes. The attachment of enzymes to carbon surfaces is achieved through the activation of the carbon surface and the chemical bonding of the enzyme. Graphitic carbon consists of giant sheets of fused aromatic rings, stacked coplanarly. An uninterrupted basal plane surface is non-ionic, of low polarity, hydrophobic and rich in π-electron density. Without alterations, the basal plane surface of carbon is somewhat barren to synthetic coupling reactions. (On the other hand, the high π-electron density is conducive to strong chemisorption interactions and could be used as a basis of adsorptive modification particularly where unsaturated compounds and aromatic rings are involved). The perimeter of the basal plane structure on the other hand is terminated as a graphitic edge plane on which chemical functionalities such as carboxylic, hydroxy, phenolic, quinone, lactone and other carbonyl containing groups abound. Any cleavage of graphite results in reactive edge plane 'dangling' valences that become satisfied by reaction with oxygen and water. Graphitic edge planes are therefore attractive surfaces

for chemical coupling and modification procedures. Edge planes are also polar and thus fairly hydrophilic implying that aqueous coupling reactions can be used with a good degree of success. Several chemical pretreatments have been proposed for enhancing synthetically useful carboxylic acid and hydroxylic groups on the edge planes. Carboxylic acid coverages on spectroscopic carbon rods, highly orientated pyrolytic graphite, glassy carbon and pyrolytic graphite can be enhanced by simply heating in air at 400–500°C. Treatment in a radio frequency (RF) O_2 plasma is an equally effective procedure. To enhance hydroxylic groupings on pyrolytic graphite surfaces, Lin and co-workers (1977) first thoroughly oxidized the surface with an O_2 RF plasma and then reduced it with ethereal $LiAlH_4$. More recently, work in this laboratory has shown that simple oxidation of graphite particles with 1 M nitric acid at room temperature can also greatly increase the surface concentration of hydroxyl functionalities (Cardosi, 1994). Hydroxyl-containing surfaces are particularly attractive because they can react with polyfunctional cyanuric chloride (trichloro-*sym*-triazine) producing an ether linkage that is chemically and electrochemically stable in organic solvents and aqueous solutions (Lin *et al.*, 1977). The attached cyanuric chloride (CC) can further react with a variety of substances including hydroxyl and amino compounds, alkyl and aryl Grignard reagents and organic hydrazine reagents. This approach has been used to attach the enzymes horseradish peroxidase and glucose oxidase to the surface of graphite particles (Cardosi and Birch, 1993; Cardosi, 1994). The synthetic scheme for the attachment of horseradish peroxidase to the graphite particles is shown in Figure 8.4. In an analogous fashion to metal oxide

---

Figure 8.4

Synthetic scheme showing the usage of cyanuric chloride in the covalent attachment of a protein to the surface of a carbon particle.

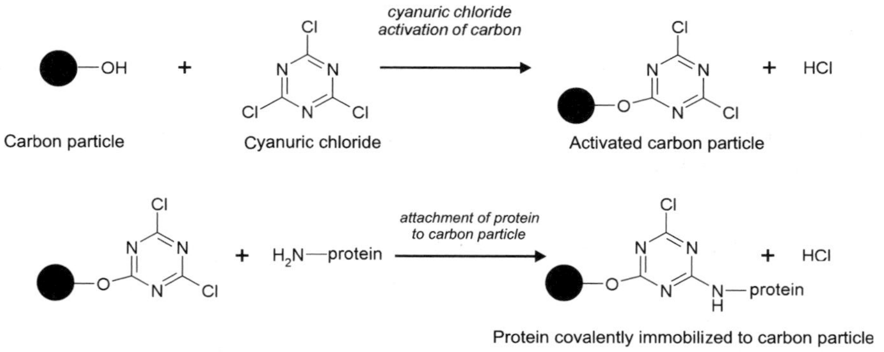

electrodes, the derivitization of the carbon surface with cyanuric chloride can be easily followed by XPS. Here, use is made of the appearance of the Cl 1s peak in the XPS spectrum following reaction with CC. It must be stressed that the actual knowledge of the absolute population and chemical reactivities of the various groupings on carbon surfaces is still limited. Although one of the treatments discussed above may appear to enhance the efficacy of a chemical coupling procedure, e.g. by producing larger electrochemical waves for attached redox species, this in itself is an indirect form of evidence which explicitly presupposes that the coupling reaction proceeds as planned.

8.3.1.4   *Immobilization in an electrochemically grown organic polymer*

Direct *in situ* formation of a polymer film from a solution of monomers can be induced electrochemically. Through electrochemical initiation, a monomer such as pyrrole or thiophene is oxidized to a polymerizable radical as illustrated in Figure 8.5.

This method is applicable to many electrode materials and also many monomer compounds. The technique is often simple to carry out, requiring only basic electrochemical instrumentation. Furthermore, the growth of the polymer film can be controlled by stepping or cycling the electrode potential (Dicks *et al.*, 1993). The properties and the reproducibility of the surface polymer ultimately depend upon the nature of the polymerization solution, the electrode material and, to some extent, the cell geometry and attendant hydrodynamic conditions. Another material that has proved particularly useful for the immobilization of enzymes at electrode surfaces is polyaniline.

If an enzyme is present in the aqueous solution containing the monomer, molecules of the enzyme will become physically trapped within the growing matrix during film growth. This approach has several significant advantages as a general method for preparing enzyme electrodes. First, the method is flexible and can be readily controlled. Second, it is simple to carry out

Figure 8.5

Scheme showing the electrochemical oxidation and subsequent polymerization of pyrrole onto the surface of an electrode.

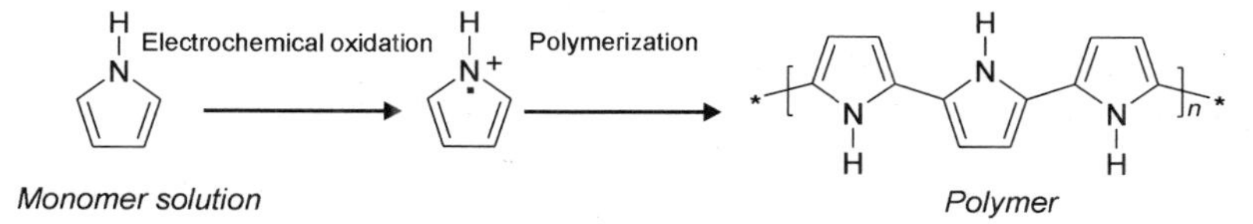

Monomer solution   *Polymer*

and usually results in enzyme loadings with high activity. Third, it is possible to co-immobilize more than one enzyme either in the same film or by growing different layers one on top of each other (note that this is only possible with conducting polymers) and finally, the polymer deposition is localized at the surface of the electrode so that the method is suited to the spatially localized deposition of enzymes onto microelectrode arrays. An example of this approach is the entrapment of glucose oxidase within a conductive matrix of polypyrrole at the surface of a platinum electrode (Foulds and Lowe, 1986; Bartlett and Whitaker, 1987a,b). Film growth occurs because the oxidation of the monomer (pyrrole) results in the formation of a highly reactive radical that reacts with neighbouring pyrrole molecules to give a polymer that is predominantly α-α' coupled (Diaz, 1981) although some branching of the polymer chains is thought to take place through β-coupling reactions. The resulting polymer has a net positive charge and incorporates anions from the bulk solution to maintain charge neutrality (Bartlett and Cooper, 1993).

Although the above example makes use of conducting polypyrrole films, conductivity is not a prerequisite for a suitable immobilization medium. Indeed, electrode films incorporating enzymes have been constructed from a wide range of materials, either conducting or insulating, which include: polyaniline polyphenols, polypyridine and cobalt tetrakis (o-aminophenyl) porphyrin.

Clearly, the successful preparation of the enzyme layer is a primary concern in the design of an enzyme electrode. Unfortunately, this still requires a certain degree of empiricism and an appreciation of the fact that what is a suitable method for one enzyme may not work as well for another. The relevant effects that immobilization can have are summarized as follows:

- The apparent activity of an enzyme can be reduced after immobilization. This in turn may be due to several factors. First, the chosen immobilization chemistry may not be optimal and may lead to modification of the active site. Even when the immobilization does not interfere with the active site, the nature of the support may produce diffusional barriers. Also, the very act of immobilization may lead to unfavourable conformational changes in the protein or reduce the conformational mobility of the enzyme. Finally, changes in the microenvironment resulting from immobilization may also lead to apparent changes in activity.
- Although the immobilization matrix is often seen as solely a support for the enzyme, it may nevertheless introduce partitioning effects. A positively charged matrix for example will exclude protons so an enzyme in this matrix will exhibit a

lower optimal pH than usual. Similarly, a hydrophobic substrate will partition into a hydrophobic matrix and thus lower the apparent K_M for that substrate.

8.4   Transducers

8.4.1   *Electrochemical transducers*

The physicochemical changes produced by the biological material must be converted into an electrical output signal by an appropriate **transducer**. On the one hand, the transducer may be unspecific and indicate general parameters such as reaction enthalpy (thermistor), mass change (piezoelectric crystal oscillator) or layer thickness (reflectometer). On the other hand, a specific indication may be achieved with ion-selective electrodes or amperometric electrodes responding to species such as hydrogen peroxide or oxygen.

There are three types of electrochemical transducers that have been used in the preparation of enzyme electrode probes: potentiometric, amperometric and conductometric. Conductometric sensors however are usually non-specific and have a poor signal-to-noise ratio (Duffy *et al.*, 1988) and as yet have not been applied to bioprocess monitoring. Consequently, these will not be discussed here but interested readers can refer to Turner *et al.* (1987) and Hall (1990) for more details.

Potentiometric transducers (ion-selective electrodes) measure the potential difference that is generated across an ion-selective membrane separating two solutions (an internal filling solution which contains the ion of interest at a fixed activity and the external sample solution). To ensure that the potential is stable with time, the measurement is made under conditions of virtually zero current flow. In addition, the potential is measured relative to a stable reference electrode of known potential. The equipment needed to make a potentiometric measurement includes an ion-selective electrode (for example pH electrode, NH_3 electrode, etc.) and a high input impedance potential measuring device.

Thermodynamic arguments (which will not be elaborated on here) tell us that the potential produced at the membrane corresponds to the free energy difference (ΔG) associated with the gradient of activity of the analyte ions in the inner and outer solutions. Ideally, an ion-selective electrode should obey the following relationship:

$$E = K + \left(2.303\frac{RT}{nF}\right)\log a_i$$

where E is the measured potential, R is the gas constant (8.314 J/K mol), T is the absolute temperature, n is the ionic charge and a_i is the activity of the specified ion. K is a constant comprising of contributions from various sources. Theoretically, the calibration graph obtained from an ion-selective electrode should have a slope of $59/n$ mV at standard temperature and pressure.

When an enzyme is immobilized close to the sensing head, the enzymic reaction with the analyte of interest generates a change in potential due to ion accumulation or depletion. For

Table 8.2　Examples of enzyme/ion-selective electrode combinations

Substance detected	Enzyme	Ion-selective electrode
Urea	Urease	pH, NH_3, CO_2
Glucose	Glucose oxidase	pH, I^-
L-Amino acids	L-Amino acid oxidase	NH_4^+, I^-
L-Tyrosine	L-Tyrosine decarboxylase	CO_2
L-Glutamine	Glutaminase	Cation
L-Glutamic acid	Glutamate dehydrogenase	Cation
L-Asparagine	Asparaginase	Cation
D-Amino acids	D-Amino acid oxidase	Cation
Penicillin	Penicillinase	pH
Amygdalin	β-Glucosidase	CN^-
Nitrate	Nitrate reductase	NH_4^+
Nitrite	Nitrite reductase	NH_3

example, if one wishes to make a potentiometric sensor for the detection of urea, the enzyme urease can be used. Urease, an enzyme isolated from the jack bean, catalyses the following reaction:

$$\begin{array}{c} H_2N \\ \diagdown \\ C=O + 2\,H_2O + H^+ \xrightarrow{\ \text{Urease}\ } HCO_3^- + NH_4^+ \\ \diagup \\ H_2N \end{array}$$

Thus, the reaction can be followed by using either an ammonia or carbon dioxide gas sensing electrode.

Examples of enzymes/ion-selective electrode combinations are given in Table 8.2.

Amperometric sensors monitor the flow of current at a working (polarized) electrode as a function of the applied potential. The potential of the electrode serves as the driving force for the electrochemical reaction, i.e. it is the controllable parameter in the experiment that causes the electroactive species present in the solution to be electrolysed (oxidized or reduced) at the surface of the electrode. As the potential of the electrode becomes more negative, the electrode becomes a stronger reductant (source of electrons). As the potential becomes more positive, so the electrode becomes a stronger oxidizing agent (electron sink). The current resulting from an oxidation, or reduction, is termed the faradaic current because it obeys Faraday's laws of electrolysis. The faradaic current is a direct measure of the rate of the electrochemical reaction taking place at the electrode surface and is dependent on two things: (i) the rate at which the species moves from the bulk solution to the electrode surface (mass transport) and (ii) the rate of electron transfer across the charged interface (charge transfer).

Oxidation involves the loss of electrons. Reduction involves the gain of electrons.

Amperometric enzyme electrodes are based upon redox enzymes such as glucose oxidase. This class of enzyme catalyses the oxidation of diverse substrates such as fatty acids, sugars, amino acids, aldehydes and phenols utilizing molecular oxygen as the electron sink. Hydrogen peroxide is produced as a by-product of the reaction which can be detected amperometrically. Alternatively, the rate of reaction can be monitored by measuring the depletion in oxygen tension in a sealed reaction vessel.

Oxygen or hydrogen peroxide based electrodes are strongly influenced by local fluctuations in oxygen tension that can result from changes in pH, temperature, ionic strength or partial pressure. In addition, hydrogen peroxide electrodes are prone to interference due to non-specific electrochemical oxidation of endogenous substances like ascorbate, urate, glutathione and cysteine. These problems have led to the development of chemically modified electrodes in which the natural electron acceptor for the enzyme reaction (dioxygen) is replaced by an artificial oxidant (mediators) like ferricinium, methylene blue, ferricyanide, tetrathiafulvalene, etc. (Cardosi and Turner, 1991). Mediators can be incorporated into the electrode by: (1) adsorption using a mediator organic solvent solution that is allowed to evaporate, (2) entrapment in or behind a polymer film, (3) covalent binding to polymers that can then be deposited onto the electrode surface, (4) mixing into a paste of graphite and mineral oil (carbon paste electrodes), (5) covalent attachment to preformed polymers on the electrode surface, (6) *in situ* formation during the electropolymerization of monomer units onto the surface of the electrode.

8.4.2 Biosensors based on thermal effects

A **mediator** is a small redox couple which takes the place of the enzyme's natural electron acceptor. Because mediators are mobile, they can interact at the active site of the enzyme, undergo electron exchange and then diffuse to the surface of the electrode where they become reoxidized.

These types of devices are based on the principle that enzymic reactions are exothermic in nature. This fact can be used to calorimetrically determine the amount of substrate converted to product during the enzyme-catalysed reaction (Mattiasson *et al.*, 1981; Mandenius and Danielsson, 1988). Molar enthalpies for enzyme-catalysed reactions range from 5 to 100 KJ mol^{-1}. Modern thermistors can be used to accurately measure temperature changes as low as 10^{-4} °C. As with enzyme electrodes, the enzyme is attached directly to the surface of the temperature transducer (thermistor) either by crosslinking or entrapment. Alternatively, the enzyme can be placed in a temperature controlled 'reactor column' and the heat of reaction measured by recording the difference in temperature between the column outlet and the column inlet. Although a general method for measuring enzyme-catalysed reactions, calorimetry suffers from two major drawbacks: (1) non-specific thermal effects resulting in overestimation of the substrate concentration, and (2) baseline drift due to heating of the measurement apparatus.

8.4.3   Biosensors based on optical effects

In sensors based on optical methods of detection, the modulation of electromagnetic radiation such as UV/vis absorption, bio- and chemiluminescence, reflectance, fluorescence and surface plasmon oscillation caused by the interaction of the biocatalyst with the target analyte is monitored optically. A key consideration in these types of device is the use of optical fibre waveguide technology. This involves the synthesis of two ideas: (1) the use of optical fibres to bring light of the appropriate wavelength from a spectrometer and back again and (2) the use of optical fibres as an immobilization support for the biocatalyst thereby allowing the reagent to be used on a continuous rather than a 'one-off' basis.

Because most optical sensors that have been developed to date respond to analytes that can also be sensed electrochemically, electrochemical sensors have provided the 'frame of reference' for evaluating the performance of optical sensors. This type of comparison should be treated with caution, however, because optical sensors are based on very different principles than electrical ones. Consequently, the relative merit of each approach will depend on both the particular analyte being measured and the operational demands of the particular application. In general terms, optical sensors offer the following advantages over their electrochemical counterparts.

- No reference electrode is required.
- Because the signal is optical it is not subject to electrical interference. This is particularly important for sensors that are operated in a particularly noisy environment such as in a fermentor vessel.
- The immobilized biocatalyst does not have to be in direct contact with the optical fibre.
- Optical devices are inherently safer than electrical devices.
- Optical sensors can be highly stable with respect to calibration and baseline drift.

Optical sensors that simultaneously respond to more than one analyte can be prepared using multiple immobilized reagents with different wavelength characteristics. For example, Scheper et al. (1994) described a biooptrode (optical biosensor) for the simultaneous detection of glucose, fructose, gluconolactone and sorbitol that could be used for on-line process monitoring. The

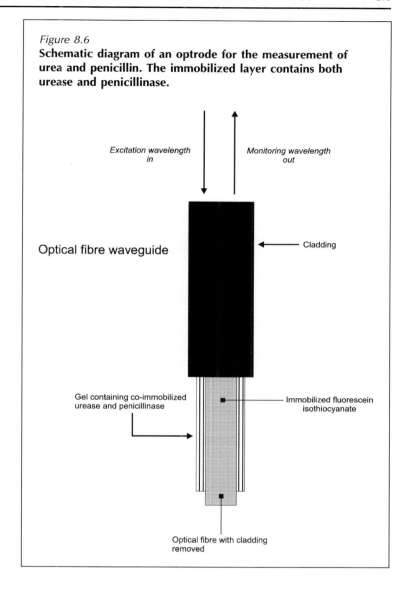

Figure 8.6
Schematic diagram of an optrode for the measurement of urea and penicillin. The immobilized layer contains both urease and penicillinase.

Excitation wavelength in

Monitoring wavelength out

Optical fibre waveguide

Cladding

Gel containing co-immobilized urease and penicillinase

Immobilized fluorescein isothiocyanate

Optical fibre with cladding removed

device made use of the glucose–fructose oxidase complex isolated from *Zymomonas mobilis*. The oxidation and reduction of the coenzyme NADP⁺/NADPH was monitored by measuring the fluorescence intensity at 450 nm. In another type of biooptrode described by the same authors (Scheper *et al.*, 1994), the pH modulated fluorescent property of fluorescein isothiocyanate was used to measure urea and penicillin after the co-immobilization of urease and penicillinase. For a diagrammatic representation of the optrode, see Figure 8.6.

8.5   Applications reported in the literature

Although perceived as an important area, there have not been many reports of biosensors for fermentation monitoring and control. This is probably due to the fact that the fermentor is not the most ideal environment for biosensors and the need to invest heavy capital towards research and development. In spite of this however, there are groups worldwide who are looking at the development and implementation of biosensors for process control. Their work, in chronological order, will be summarized in this section.

8.5.1   *Electrochemical probes*

8.5.1.1   *Potentiometric biosensors*

The first successful description of potentiometric biosensors used for process monitoring and control was described by Mizutani *et al.* (1987). These workers described an on-line fully automated system for the monitoring and control of glucose (carbon source) concentration in a fed-batch yeast fermentation system. Briefly, cell-free filtrate was withdrawn aseptically from the fermentor using a ceramic membrane cartridge. The filtrate was then passed over a potentiometric glucose biosensor and the glucose concentration determined. The resolution of the analysing system was 1 minute. Using this approach and an 'on–off' control strategy, good feedback control of the glucose concentration in the fermentor was achieved.

In 1993 Meyerhoff and co-workers utilized a flow injection analysis (FIA) system to monitor glutamine concentration in hybridoma cell cultures. Here, the enzyme glutaminase was immobilized in a reactor column through which samples of filtered growth medium were injected and the concentration of glutamine was determined from the ammonium ions produced in the flow-through glutaminase column.

Finally, in 1991 Brand *et al.* described a multi-analyte field effect transistor (FET) capable of the simultaneous detection of glucose, urea, penicillin G and cephalosporin in an FIA set-up. This was used successfully to monitor these analytes during the cultivation of a variety of microorganisms including *E. coli* K12 MF, *E. coli* JM103, *S. cerivisiae* and *Candida boidinii*.

8.5.1.2   *Amperometric biosensors*

The first reports of amperometric biosensors for process monitoring and control can be traced back to the late 1980s. For example, Romette and Cooney (1987) described a biosensor for monitoring glutamine in mammalian cell cultures. Glutaminase and glutamate oxidase were insolubilized in a thin gelatine membrane that was subsequently fixed over the surface of a polarographic oxygen electrode. Correlation between oxygen consumption and glutamine concentration provided a linear response between 0.2 and 2 mM glutamine.

In 1991, Kleman *et al.* reported a computer-assisted on-line glucose analyser for feedback control of *E. coli* cell growth in fed-batch cultures. Glucose concentration was determined after cross-flow removal of particulate matter using a commercially available glucose analyser (YSI glucose analyser). The concentration of glucose in the fermentor was maintained at a stat level of 1.5 ± 0.5 g/l by feedback control. The device was very stable with the enzyme membrane requiring changing after every 600 to 800 measurements. Holst and co-workers also used a commercial glucose analyser for the monitoring of glucose concentrations in cultures of yeast and *E. coli*. Samples were taken every 90 s and measured over the working range 0–5 g/l. This analyser, linked to an automatic injection system, was used to monitor and control glucose concentrations during continuous cultivation of *Scizosaccharomyces pombe* as well as continuous and fed-batch cultures of *S. cerivisiae*. On-line sampling was carried out through a sample loop system in the fermentor vessel itself.

A different approach was employed by Brooks *et al.* (1987) who described a dimethylferrocene-mediated glucose sensor for on-line monitoring of glucose. Despite problems of poor stability and drift with the prototype device, Bradley and Schmid constructed an *in situ* ferrocene-mediated biosensor for glucose consisting of an outer housing of stainless steel closed at one end by a sterilizable polycarbonate membrane. After sterilization of the membrane the biosensor, which consisted of four graphite electrodes press fitted into cavities on its surface, was introduced and held in place against the polycarbonate membrane. The biosensor performed satisfactorily during 4 h of continuous use and had a dynamic range up to 20 g/l^{-1} of glucose. Careful placement in the fermentor vessel ensured that fouling of the surface was minimized.

In 1990, Rishpon *et al.* described a glucose probe for *in situ* monitoring of glucose concentration during fermentation. The probe had an interesting design in that it consisted of a platinum mesh electrode (onto which glucose oxidase was immobilized) sandwiched between a dialysis membrane on one side and an oxygen-permeable silicone membrane on the other. The function of the dialysis membrane was to prevent fouling from large macromolecules and cells whereas the silicone membrane provided oxygen for the glucose oxidase reaction. This electrode was applied to the continuous monitoring of *S. cerivisiae* cultivation for a period of 28 h. During this time, however, it was noted by the authors that the probe suffered from clogging of the dialysis membrane resulting in unstable outputs.

The use of amperometric oxygen detection was used by Scheper *et al.* (1991) for the on-line monitoring and control of monosaccharides, disaccharides, lactate and amino acids during the production of alkaline protease and penicillin. Immobilized reactor columns of glucose oxidase, lactate oxidase and L-amino acid oxidase were coupled to oxygen electrodes to monitor the respective analytes. Sucrose, lactose and maltose were also monitored by first converting them to glucose using in-line reactor columns containing the enzymes invertase, lactase or maltase.

Finally, a number of groups worldwide have investigated the application of hydrogen peroxide sensors for glucose and glutamine analysis in mammalian cell culture. Renneberg and co-workers for example developed an enzyme sensor based FIA system to monitor glucose, glutamine and lactate in cultures of hybridoma cells producing Mab and interleukin 2. Cattaneo and Luong (1993) have also developed a biosensor system for monitoring glutamine concentration in mammalian cell cultures. The system consisted of two electrodes. The first had co-immobilized glutamate oxidase and glutaminase and the second incorporated glutamate oxidase alone. The first electrode responded to both glutamine and glutamate whereas the second electrode responded only to glutamate. The difference reading was used to give a measure of the total glutamine present in the culture.

An alternative method to alleviate the problem of endogenous glutamate interference is to use ion exchange. Here, an anion-exchange column is placed upstream of the enzyme columns in the biosensor–FIA system. The purpose of the anion-exchange column is to remove the negatively charged glutamate ion from the sample slug. This approach is effective not only in removing glutamate but also in removing other potential interferents like ascorbate and urate.

8.5.2 *Thermal biosensors*

The approach to process monitoring and control using thermal biosensors as the analytical device is essentially the same as that described for electrochemical sensors with the important difference that in this case, enthalpy changes are measured as opposed to electrical ones. Usage of these devices either as *in situ* devices or in an FIA manifold is however common.

In 1981 Mattiasson *et al.* described the operation of a thermistor type biosensor for monitoring glucose in untreated, unfiltered fermentation samples. Glucose oxidase was immobilized onto glass beads which were packed into a column. A dialysis membrane was used to prevent fouling.

Hundeck and co-workers (1990) described a four enzyme column system using a thermistor with a temperature resolution of 10^{-5} K. Four different enzyme columns and a blank column were used in conjunction with a calorimeter device. Glucose, fructose, maltose and sucrose were monitored simultaneously in filtered bioreactor liquor.

Finally, in 1993 Rank *et al.* described an enzyme–thermistor biosensor system using split flow for monitoring penicillin V, glucose and ethanol in bioreactor vessels. Non-specific heat effects (for example from the movement of the flowing stream) were eliminated by the use of a reference column.

8.5.3 *Optical biosensors*

Once again, the ethos of developing sensors based on optical transduction is essentially the same, i.e. an enzyme such as glucose oxidase is used to give the specificity and the enzymic reaction is monitored through an optical change. For example, Huang *et al.* (1991) described a chemiluminescent detection scheme linked to FIA for on-line monitoring of glucose in animal cell cultures. Here, glucose was converted with immobilized glucose oxidase and the hydrogen peroxide produced reacted with luminol and the intensity of the emitted light at 425 nm was measured. By using immobilized glutaminase and glutamate oxidase, Cattaneo and Luong (1993) used the same methodology to monitor glutamine concentration in animal cell culture.

Fibre-optic based oxygen sensors have also been used in conjunction with glucose oxidase. Oxygen optrodes are generally based on fibre-optics covered with a rubber film containing decacyclene, a fluorescent dye which responds strongly to oxygen partial pressure. Filtered samples from fermentor vessels can then be passed through immobilized reactor columns and the decrease in oxygen partial pressure detected using the optrode downstream.

Reduced nicotinamide coenzymes fluoresce strongly. Thus, glucose and lactate can be detected by passing samples through immobilized reactor columns containing the enzymes such as glucose, alcohol and lactate dehydrogenase. NAD^+ is provided in the sample buffer and the NADH formed in the biocatalytic reactions is detected via fluorescence. This approach has been investigated by Ogbomo and co-workers for the parallel determination of glucose and ethanol during baker's yeast fermentation. Finally, the fluorescence properties of NADH have been exploited to monitor ethanol, glucose, gluconolactone and sorbitol in fermentation vessels.

Summary

As this chapter has indicated, there is certainly a great deal of interest in producing measurement methodologies incorporating biosensors for process monitoring and control. It is well worth reiterating the reasons why:

- Biosensors offer simple and specific measuring devices which can be used *in situ* or off-line.

- The response times of biosensors are sufficiently fast to allow real-time information gathering and processing.

- Biosensors can be prepared for a wide range of analytes.

- There is still sufficient interest in the research community to ensure that significant advances will be made in the field.

- Reliable techniques are now available for the mass manufacture of biosensor devices.

- Biological sensing elements can be coupled to a variety of transducers with the result that the final sensor configuration can be tailored to meet a particular set of operational parameters.

This apparent surge of optimism must however be tempered with some caution because there are still a number of important failings which must be overcome. The most important of these is the inherent fragility of the biological element. This has the result of producing a sensor with variable operational characteristics (drift, sensitivity) as a function of time. Work in this area, both theoretical and practical, continues apace. The most important development in this area is, in my view at least, the development of artificial bio-mimetic systems.

Reviewing the literature reveals that FIA is often the technique of choice for integrating the biosensor with the fermentation process. This approach is acceptable for 'slow' systems such as mammalian cell growth but may not be so well suited to faster microbial systems which demand faster sample throughput.

Multiple analyte monitoring is another important requirement for effective bioprocess control. Because of the range of biological elements available, biosensors should make an important contribution in this area. This is particularly true if biosensor technology is combined with microfabrication and chip technology. The multianalyte chip which is capable of biosensing 100 different analytes is certainly an attractive research and commercial goal.

References

Alvarez-Icaza, M. and Schmid, R. (1994) Observations of direct electron transfer from the active centre of glucose oxidase to a graphite electrode achieved through the use of mild immobilization, *Bioelectrochem. Bioenergetics*, **33**, 191–199.

Bartlett, P.N. and Cooper, J.M. (1993) A review of the immobilization of enzymes in electropolymerised films, *J. Electroanal. Chem.*, **362**, 1–12.

Bartlett, P.N. and Whitaker, R.G. (1987a) Electrochemical immobilization of enzymes, Part I. Theory, *J. Electroanal. Chem.*, **224**, 27–35.

Bartlett, P.N. and Whitaker, R.G. (1987b) Electrochemical immobilization of enzymes, Part II. Immobilization of glucose oxidase in thin films of electropolymerised phenols, *J. Electroanal. Chem.*, **224**, 37–48.

Bradley, J. and Schmid, R.D. (1991) Optimization of a biosensor for *in situ* fermentation monitoring of glucose concentration, *Biosensors and Bioelectronics*, **6**, 669–674.

Brand, U., Reinhardt, B., Rüther, F., Scheper, T. and Schügerl, K. (1991) Bio-field-effect transistors for process control in biotechnology, *Sensors and Actuators*, **4**, 315–318.

Brooks, S.L., Ashby, R.E., Turner, A.P.F., Calder, M.R. and Clarke, D.J. (1987) Development of an on-line glucose sensor for fermentation monitoring, *Biosensors*, **3**, 45–56.

Cardosi, M.F. (1994) Hydrogen peroxide sensitive electrode based on horseradish peroxidase modified platinised carbon, *Electroanalysis*, **6**, 89–96.

Cardosi, M.F. and Birch, S.W. (1993) Screen printed glucose electrodes based on platinised carbon particles and glucose oxidase, *Anal. Chim. Acta*, **276**, 69–74.

Cardosi, M.F. and Turner, A.P.F. (1991) Mediated electrochemistry: a practical solution to biosensing, in Turner, A.P.F. (ed.) *Advances in Biosensors*, Vol. 1, London: JAI Press, pp. 125–169.

Cattaneo, M.V. and Luong, J.H.T. (1993) Monitoring glutamine in animal cell cultures using a chemiluminescence fiber optic biosensor, *Biotechnol. Bioeng.*, **41**, 659–665.

Diaz, A. (1981) Electrochemical preparation and characterisation of conducting polymers. *Chem. Scripta*, **17**, 145–148.

Dicks, J.M., Cardosi, M.F. and Turner, A.P.F. (1993) The application of ferrocene modified n-type silicon in glucose biosensors, *Electroanalysis*, **5**, 1–9.

Duffy, P., Saad, I. and Wallach, J.M. (1988) New developments of conductometric measurements in analytical biochemistry, *Anal. Chim. Acta*, **2**, 267–272.

Foulds, N.C. and Lowe, C.R. (1986) Immobilization of glucose oxidase in ferrocene-modified pyrrole polymers, *Anal. Chem.*, **60**, 2473–2478.

Hall, E. (1990) *Biosensors*, Milton Keynes: Open University Press.

Huang, Y.L., Li, S.Y., Bremel, B.A.A., Bilitewski, U. and Schmid, R.D. (1991) On-line determination of glucose concentration in animal cell cultures based on chemiluminescent detection of hydrogen peroxide coupled with flow injection analysis, *J. Biotechnol.*, **18**, 161–166.

Hundeck, H.-G., Sauerbrei, A., Hübner, U., Sceper, T., Mandenius, C.F., Koch, R. and Antranikian, G. (1990) Four channel enzyme thermistor system for process monitoring and control in biotechnology, *Anal. Chim. Acta*, **238**, 211–221.

Kleman, G.L., Chalmers, J.J., Luli, G.W. and Strohl, W.R. (1991) A predictive and feedback-control algorithm maintains a constant glucose concentration in fed-batch fermentations, *Appl. Environ. Microbiol.*, **57**, 910–917.

Lin, A.W.C., Yeh, P., Yacynych, A.M. and Kuwana, T. (1977) Cyanuric chloride as a general linking agent for the attachment of redox groups to pyrolytic graphite and metal oxide electrodes, *J. Electroanal. Chem.*, **84**, 411–419.

Mandenius, C.F. and Danielsson, B. (1988) Enzyme thermistors for process monitoring and control, *Methods Enzymol.*, **137**, 307–322.

Mattiasson, B., Danielsson, B., Mandenius, C.F. and Winquist, F. (1981) Enzyme thermistors for process control, *Ann NY Acad. Sci.*, **369**, 295–305.

Meyerhoff, M., Trojanowicz, M. and Palsson, B.O. (1993) Simultaneous enzymatic/electrochemical determination of glucose and L-glutamine in hybridoma media by flow injection analysis, *Biotechnol. Bioeng.*, **41**, 961–975.

Mizutani, S., Iijima, S., Morikawa, M., Shimizu, K., Matsubara, M., Ogawa, Y., Izumi, R., Matsumoto, K. and Kobayashi, T. (1987) On-line control of glucose concentration using an automatic glucose analyser. *J. Ferment. Technol.*, **65**, 325–331.

Murray, R.W. (1980) Chemically modified electrodes, *Acc. Chem. Res.*, **13**, 135–141.

Ogbomo, I., Kittsteinereberle, R., Englbrecht, U., Prinzing, U., Danzer, J. and Schmid, R.D. (1991) Flow-injection systems for the determination of oxidoreductase substrates – applications in food quality control and process monitoring, *Anal. Chim. Acta*, **249**, 137–143.

Rank, M., Gram, J. and Dannielsson, B. (1993) Industrial on line monitoring of penicillin V, glucose and ethanol using a split flow modified thermal biosensor. *Anal. Chim. Acta*, **281**, 521–526.

Rishpon, J., Shabtai, Y., Rosen, I., Zibenberg, Y., Tor, R. and Freeman, A. (1990) *In situ* glucose monitoring in fermentation broth by 'sandwiched' glucose-oxidase electrode (SGE), *Biotechnol. Bioeng.*, **35**, 103–107.

Romette, J.L. and Cooney, C.L. (1987) Glutamine electrode for on-line mammalian cell culture process control, *Anal. Lett.*, **20**, 1069–1081.

Ruzicka, J. and Hansen, E.H. (1988) *Flow Injection Analysis*, 2nd edn, New York: Wiley.

Scheper, T., Müller, C., Anders, K.D., Eberhardt, F., Plötz, F., Schelp, C., Thorsden, O. and Schögrel, K. (1994) Optical sensors for biotechnological applications, *Biosensors Bioelectron.*, **9**, 73–83.

Scheper, T., Brandes, W., Grau, C., Hundeck, H.G., Reinhardt, B., Ruther, F., Plotz, F., Schelp, C., Schugerl, K., Schneider, K.H., Gifforn, F., Rehr, B. and Sahm, H. (1991) Applications of biosensor systems for bioprocess monitoring, *Anal. Chim. Acta*, **249**(1), 25–34.

Schmidt, H. (1993) Biosensors and flow injection analysis in bioprocess control, *J. Biotechnol.*, **31**, v.

Turner, A.P.F., Karube, I., and Wilson, G.S. (eds) (1987) *Biosensors: Fundamentals and Applications*, Oxford: Oxford University Press.

9 Control of Fermentations: An Industrial Perspective

Craig J.L. Gershater

9.1 Requirement for control

Control is generally defined as the power to direct or influence. In the case of fermentation control, the requirement to control a given biotechnological process is generally dictated by the need to bring about a desired outcome such as maximizing output of product formation. Microorganisms used in industrial processes have been isolated by virtue of their ability to overproduce certain commercially significant attributes. These attributes can range from the ability to produce carbon dioxide for leavening activity in the case of *Saccharomyces cerevisiae* (baker's yeast) to the production of useful microbial metabolites such as alcohol from yeast or antibiotics from filamentous bacteria.

Whatever the reason for commercial exploitation, the microorganism will have been obtained directly or indirectly from an ecosystem very remote from that of the fermentation laboratory and process plant. The metabolic attribute to be exploited will have evolved as a result of environmental factors unknown to the fermentation scientist and therefore process optimization must seek to replicate those environmental factors responsible for expression of the desired attribute in the totally 'artificial' environment of the industrial bioreactor or fermentor.

9.1.1 *Microbial growth*

The principal fermentation control requirement is population growth (see Chapter 3 for more details) of the microorganism of interest. In its natural habitat, the microbe will respond to environmental stimuli such as excess nutrients by synthesizing enzymes and biomass capable of exploiting the resource as effectively as possible. In the fermentor system, the microbe will be inoculated into the fermentation medium 'feast', and will thus attempt to colonize this environment through rapid growth.

The role of the fermentation control strategy is to provide, by control of environmental effectors such as temperature, aeration, pH and dissolved oxygen, the optimum conditions for growth and 'colonization'. The fermentation scientist uses an environment that is capable of being controlled to a limited degree, i.e. the fermentor, to develop a control strategy that will modify inputs to the fermentor system in order to achieve the desired outputs. This system of modifying inputs to obtain a desired output from a control system is often described as a **control loop**.

9.1.2 *Nature of control*

A fermentation development programme seeks to establish what control set points are needed for the control loops in the control system. It is easy to forget that the control system can be human! There are many examples of accurate human control loop systems including the motor car. In the motor car the driver observes the speed of the vehicle (the output from the system) by looking at the speedometer (sensor). If the car is travelling too fast, the control system (the driver) reacts by reducing pressure on the accelerator thus reducing the flow of fuel to the engine (the input to the system). The car slows and the driver observes whether the speed matches the desired outcome (speed limit); if it does, the adjustment to the accelerator will be modified again, and so on. Control of a fermentation system can similarly be by manual intervention of control valves adjusted as a result of changes observed on gauges, however this chapter will concentrate on the elements of automatic (computer) control.

9.1.3 *Control loop strategy*

The basic element of a control system is the control loop. In Figure 9.1 the various components that make up the control

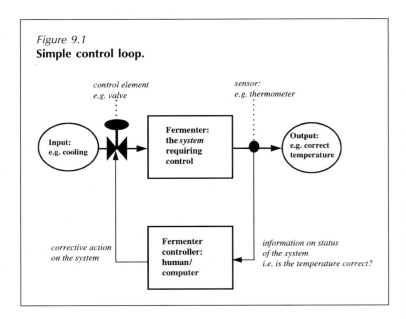

Figure 9.1
Simple control loop.

loop are summarized. At the centre of the control loop is the system requiring control, the fermentor. The system can be affected by a number of influences and in this case it is the temperature of the system that is to be controlled. A very simple control loop is shown where only cooling may be applied to the fermentor principally to remove metabolic heat during incubation. The temperature of the fermentor is measured using a thermometer and either a human operator or a control system can monitor this measurement and make adjustments if the desired temperature (the set point) and the actual temperature are not equal (error in the system). If the temperature of the fermentor is too high then the flow of cooling water can be increased (input to the system) by opening the valve controlling water to a heat exchanger. The temperature of the fermentor system will decrease and a new temperature measurement taken, if the temperature is correct, i.e. the actual temperature equals the set point then the controller can close the water control valve and the temperature allowed to stabilize. In reality dynamic fermentor systems never fully stabilize and constant control is required, hence the advantage of automated systems which maintain a constant vigil on the process and adjust inputs to the system as soon as required.

9.2   Sensors

Development of fermentation processes over the past 50 years has been accomplished in large scale reactors (> 10 litres) using relatively few sensors. The problem with monitoring the fermentation system is maintaining the sterile integrity of the fermentor hence any sensor has to be capable of sterilization either *in situ* during steam sterilization or remotely and then aseptically introduced to the fermentor.

9.2.1   *History*

In the 1940s the sensors were largely based on manual sampling and off-line analysis, control was improved through the 1950s and 60s by the use of limited electrical signals controlling pneumatic outputs (valves, etc.). In the 1970s new sensors capable of being sterilized were introduced including pH and dissolved oxygen probes. In addition more accurate methods of measuring flow rates (both liquid and gas) became available and other engineering parameters such as motor power to the agitator and enhanced nutrient feed addition systems were developed. As well as improvements to sensors and other measuring devices minicomputers were introduced to provide simple control and data logging.

Since the 1970s and 80s relatively few new sensors (see Chapter 8 for more details) have emerged for on-line analysis of fermentation parameters. Recent techniques that have become available include biomass probes, on-line liquid chromatography systems, near infra-red spectroscopy, and so on. For the most part, although these systems do address more complex issues related to monitoring the fermentation, the reliability and relevance of some of these measurements together with prohibitive cost tend to preclude these systems from everyday fermentation development programmes.

9.2.2 *Typical fermentation sensors*

The control elements (sensors) that should be considered routine for most (aerobic) fermentation systems are:

1. Temperature: measured using a platinum resistance thermometer (PRT probe) where the increase in temperature is proportional to the increase in electrical resistance in the probe.
    *   Temperature will be controlled by the addition of cooling water to a jacket or cooling finger of a fermentor; heat will be added by direct heating of the vessel or its contents (electrical heating mantle or 'hot finger') or by the injection of hot water or steam to the circulating water in a jacket or heat exchanger.
2. Air flow rate: measured using a standard pressure drop device such as variable area flowmeters or more often a mass flow sensor.
    *   Air flow will generally be controlled using a proportional (0 to 100% open) valve upstream of the sterile inlet filter on a fermentor.
    *   Air flow is frequently quoted as VVM, volume of gas per volume of liquid per minute, fermentor design generally permits up to 2 VVM.
3. Vessel pressure: measured using diaphragm protected Bourdon gauges or strain gauge pressure transducers.
    *   Pressure in the vessel is induced during *in situ* steam sterilization and during normal incubation with the introduction of air. Control of vessel pressure is by regulation of the vent gas from a fermentor.
    *   Pressure is generally a negatively acting loop in that a fully driven output valve (100% open) results in minimal pressure in the fermentor.
    *   The units of pressure are generally bar gauge, i.e. pressure within the reactor above atmospheric.
4. Vessel agitation rate: measured using proximity detectors to detect the speed of the shaft.
    *   Impellors or mixers are controlled by standard motor controllers and the units are revolutions per minute.

- Agitation power is sometimes measured using current transformers measuring the electrical power consumption of the motor, the units of power are Watts.

5. pH: measured using steam sterilizable combined glass electrodes.

   pH is controlled by the use of buffers added into the medium or by the addition of acids or base titrants.

6. Dissolved oxygen: measured using polarographic type probes, here galvanic voltages on a membrane-covered oxygen reducing cathode induce a current (amperometric) proportional to the amount of oxygen diffusing through the membrane.

   - Dissolved oxygen is controlled by altering the status of those control loops affecting dissolved oxygen generation. An increase in dissolved oxygen is induced in fermentation medium by:
     1. increasing the air flow rate (volumetric increase in oxygen),
     2. increasing agitator speed (smaller air bubbles increasing the surface area available for diffusion),
     3. increasing over pressure in the fermentor (generally increasing the residence time for air bubbles).
   - Dissolved oxygen is measured as the partial pressure of oxygen at the electrode surface and hence is quoted in percentage saturation (often referred to as dissolved oxygen tension, DOT). To obtain a mass, a fully saturated solution of oxygen in water is approximately equivalent to 1.2 mM/litre.

7. Foam: detected by observation, or conductance or capacitance probes completing an electric circuit when foam is contacted.

   - Foam is controlled by reducing the cause of foaming, i.e. high aeration and/or agitation rates or by the addition of antifoaming agents.

Figure 9.2 shows the main control elements of a fermentor, there are many different configurations which may be specified but the one shown would be capable of providing data and control options for an aerobic fermentation.

9.2.3 Control action

- Temperature control would be achieved by the regulated supply of temperature controlled water to the jacket of the fermentor. In the case illustrated there are two thermometers in the system, one indicating the temperature of the medium/broth, the other indicating the temperature of the return water flow. In this configuration the two signals may be

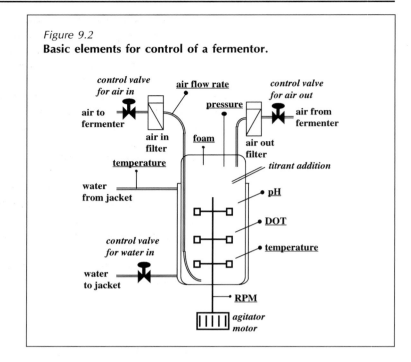

Figure 9.2
Basic elements for control of a fermentor.

compared and control options could include linking their function to provide a regulated flow of temperature controlled water to the jacket via the control valve shown.

- Airflow control would be achieved by regulating the linear gas flow rate to the system by adjusting the airflow control valve shown in response to signals coming from the air flow sensor.
- Pressure control is achieved by regulating the flow of off gas from the fermentor. It is probably beneficial to have the airflow and pressure control functions independent of each other. For the most part these control loops will act to the limit of their engineering configuration and if a particular set point, say for airflow, cannot be achieved because of excess backpressure then this may be noted as a process constraint and the operating protocol adjusted accordingly.
- Agitator speed is controlled by direct feedback of the revolutions per minute using a suitable tachometer.
- pH is generally controlled by the addition of acid or base titrant in response to changes in pH value during the fermentation. One important aspect of the pH loop is the calibration of the probe before and after sterilization. This is achieved before sterilization by the adjustment of slope and intercept values on the pH controller to be used during the fermentation. Subsequent to sterilization it is possible to check the calibration of the pH probe by comparing the pH of a

sample of medium withdrawn from the fermentor using a calibrated external pH meter. The function of the pH control loop will be discussed in more detail later.

- Dissolved oxygen control is achieved in a number of different ways. It is possible to rely totally on increasing agitator speed to shear air bubbles rising through the broth to increase surface area and thus mass transfer of oxygen from the gaseous to liquid phase. In addition however it is also possible to cascade the control output from the dissolved oxygen controller to agitator speed, airflow and pressure or any combination of these. In a particularly high oxygen demanding fermentation all three outputs may be specified, care has to exercised in the use of independent control loops such as agitator speed that this loop will be the 'slave' to the DOT 'master' control loop function. The function of the DOT control loop will be discussed in more detail later.
- Foam is controlled by the addition of antifoam agents via the feed addition system in response to a contact probe detecting rising foam.

9.3   Controllers

Control instrumentation has developed rapidly in recent years and the range of options available for fermentation control is extensive (see Chapter 10 for more details). The development of integrated circuits in the last 20 years or so has meant that complex control functions can be devolved to cheaper instruments, or conversely more sophisticated control options become increasingly available to the fermentation scientist.

9.3.1   Types of control

There are two fundamental types of control which may be incorporated into a fermentation system: sequence control and loop control. Sequence control is that part of control which permits automation of the fermentor operation such as sterilization and other valve automation sequences, i.e. for the most part providing an ordered array of digital (on/off) signal control. Loop control is generally associated with that part of control dealing with combinations of digital and analogue control signals and although used in sequence control (particularly automated sterilization sequences) is most often associated with control of incubation. The industrial fermentation scientist will frequently wish to adapt the control strategy for optimal process performance and in order to ensure versatility, bespoke control programs may be specified although the costs and potential delays associated with this approach are likely to be significant.

Control of fermentation systems can be achieved by the use of discrete single loop controllers, by programmable logic controllers controlling both sequence and loop functions and by the use of specific software packages controlling all aspects of the fermentation. This type of control is sometimes called Distributed Digital Control (DDC) and defines boundaries of operation whereby whole plant control can be achieved using computer-based systems.

9.3.2   *Control algorithms*

Whatever type of controller is selected for fermentation control, effective action will depend on the response of the controller. This response is determined by the nature of the control algorithms programmed into the system. A control algorithm is a mathematical representation of the steps required to achieve effective control, most often programmed as equations in which the controller output is a function of signal deviation away from a set point. As indicated in Figure 9.1, most controllers will function as feedback control systems, i.e. the deviation of measured variable compared with the desired set point will determine how large the control effect should be on the fermentor system via the appropriate control algorithm.

9.3.3   *PID*

The types of control algorithm most frequently encountered are 3-term or PID controllers. The PID controller is made up of three elements P: proportional, I: integral, D: derivative/differential; the purpose of these functions is to provide a fast-acting response to process deviation and scale the response to the output to achieve smooth control action. The characteristics of PID control are:

- **Proportional control** provides an output, the magnitude of which is proportional to the deviation between measured variable and set point.
- **Integral control** tends to reduce the effect of proportional control alone helping to bring the measured variable back to set point faster by minimizing the integral of control error.
- Derivative action also tends to reduce the effect of proportional control alone, this time by estimating the slope of measured variable with time and maximizing the slope of the measured variable compared with the set point.

The effect of PID control on an uncontrolled variable is shown in Figure 9.3. As indicated the algorithm tends to 'drag' the analogue value towards the set point, the speed and effectiveness of this control response is a function of the parameters

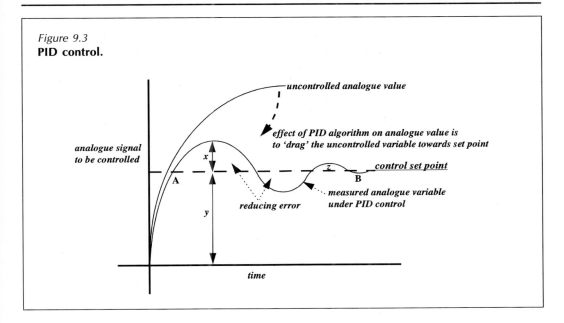

Figure 9.3
PID control.

specified by the PID algorithm, the process of setting these parameters is termed, tuning the loop. The method is generally carried out by trial and error with the loop on-line. There are three parameters associated with PID loop tuning, these are the proportional band (usually set as a percentage of full-scale deflection of the analogue signal), the integral time, and the derivative time. The method chosen for loop tuning will be recommended in control instrument vendors' instructions, the sequence generally follows:

1. Set the proportional band first with no integral (maximum value) or derivative function (minimum value).
2. When oscillations occur with small process perturbations decrease the integral time (from its maximum) until oscillations reoccur, then reset the integral time to minimize oscillations.
3. Establish whether control is adequate with PI control alone, if not satisfactory following process perturbations, increase the derivative time constant (from its minimum) to minimize oscillations.

PID loop performance can be assessed by measuring the initial overshoot as a ratio of $x : y$. A measure of PID loop effectiveness can be assessed by examining the control error decay ratio of $z : x$, in addition, the 'gain' of the controller, i.e. how fast the measured variable attains the set point when full control is applied is described by point A. Point B describes the 'settling time' for the loop, i.e. when the measured variable is within a fixed percentage of the set point value, e.g. ±5%.

9.4   Design of a fermentation control system

The first stage in designing a fermentation control system is to clearly establish what the process objectives are for the automation project.

9.4.1   Control system objectives

Objectives may be typically divided into various 'categories' of control, for example:

1. Control of basic incubation functions only, typically air-flow, agitator speed and temperature.
2. Automation of the control of incubation only relying on manual operation of valves associated with sterilization.
3. Full automation of the fermentor system including all sterilization and auxiliary vessel control.
4. Advanced control options including event based control.
5. Advanced computing methods for inferential control.

Whatever type of controller is considered necessary, the process scientist will have to specify the type of control and then make this specification available to the manufacturer of various control systems. The more complex the fermentor system to be controlled the more detailed the specification will need to be.

1. **Basic control** will typically be achieved using single loop controllers and is frequently associated with autoclavable fermentors. These controllers tend to combine both amplification of process signals and control function in one 'box', this functionality may also include a local readout function in engineering units, i.e. values recognizable to the operator. There is a wide range of commercially available single loop controllers some of which have been developed with control of fermentation processes in mind. Configuration of the controllers will follow manufacturers' guidelines, but as with all control options, factors such as signal type (current or voltage) and output device will have to be considered before final control specifications can be issued to potential vendors.
2. **Incubation control** may include other features including operation of feed delivery systems and control of more complex loops such as dissolved oxygen by multiple outputs which may be specified by the operator. The specification in this case may be more complex and be subject to negotiation with the instrument vendor. Sterilization of this type of vessel may be by autoclaving or by *in situ* steam sterilization but this will be done manually, probably to minimize cost.

3. **Full incubation and sterilization control** of the fermentor will require a control function for valve sequencing (digital control) as well as control of analogue parameters for incubation (process variables and control of proportional valves). The requirement for valve sequencing as well as more complex patterns of analogue control is met by the use of Programmable Logic Controllers (PLC) or often fermentor manufacturer's own control software running on a personal computer. This type of control is generally required for larger fermentors (> 20 litres) or where a number of identical vessels are to be purchased and manual sterilization will be too manpower intensive.

    The need for a detailed specification is determined by whether the vendor's own software meets the needs of the fermentation scientist, if it does then the basic requirements of sterilization and incubation control may be easily identified from the vendor's own specification. If 'however' pre-existing software does not meet the needs of the scientist then a much more complex and detailed specification may be required because it is most likely that the software will have to be written specifically for the purpose. It is in the best interests of both customer and vendor to agree the objectives of the bespoke software prior to code being written, costs and timescales can very quickly escalate out of control. The benefit of using vendor's own software is speed of implementation and probably lower costs, the disadvantages are that you get what the vendor thinks you want, not what you might actually need. The opposite is generally the case with PLCs, costs and timescales will be higher but if you get the specification right you get exactly what you want and need.

4. **Advanced incubation control** regimes may be required where complex fermentation patterns are required and changes to set points may be needed on-line, particularly in response to specified 'events'. This will be discussed in more detail later. The need for a detailed specification is very high under these circumstances, and clear milestones and checkpoints must be agreed during contract negotiations. It is quite likely that stage payments will be part of a contract. Large fermentor control tasks may be related to both complexity of control and numbers of fermentors within a system. If the task involves large numbers of fermentors to be controlled then a system of distributed control will probably be required. The architecture for distributed control is shown in Figure 9.4.

    Figure 9.4 illustrates one possible control option for the control of four fermentors. In this configuration each fermentor is independent of the next and control is effected by each fermentor being equipped with its own controller

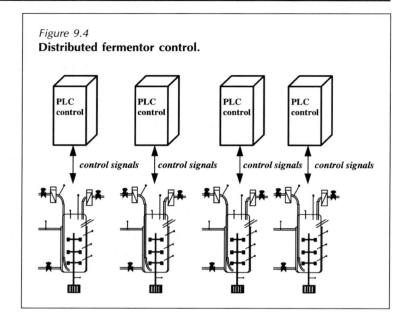

Figure 9.4
Distributed fermentor control.

(in this case a PLC). This type of distributed control has the advantage that failure of a control system will only affect one fermentor, the experimental design may be ruined by such an occurrence but without a distributed system, failure of a single controller will result in failure of the entire run. The disadvantage of distributed control is that the cost may be prohibitive to implement one controller per vessel. In the author's laboratory a compromise has been reached with vessels < 100 litres working volume where a single PLC controls a pair of identical fermentors.

5. **Advanced computing** methods may be required where analytical systems are inadequate to optimize the fermentation process and cost of goods or value of product warrant investment in 'next generation' computing methods. Inference methods are available which algorithmic sensors can function to estimate for analytes and processes for which no other sensor or probe exists. The sort of computing methods being developed include expert systems, artificial neural networks, and model-based systems. These self-learning or decision-making systems rely on process pattern recognition to identify regions of the fermentation process more or less susceptible to perturbations. Once identified, remedial action beyond the scope of experienced operators or less sophisticated controllers can be initiated to reduce cost, avoid catastrophes, or simply improve the process. These systems can often only be justified in production environments where the value of the product warrants large investment in time

and money to maximize productivity. As these methods become more widely available the costs of implementation of advanced computing methods will decrease and wider applications will be sought and introduced.

9.4.2   Fermentation computer control system architecture

In defining the overall strategy for computer control of fermentations, a system architecture will emerge to indicate how the system will be operated, how data will flow around the system and how that data can be captured, stored and interrogated when required. The control system when configured is an information exchange system.

Information about the progress of the fermentation is detected by sensors, transmitted to amplification/signal conditioning units that forward this information to the controller. The controller receives the information, compares it to pre-existing information about how the fermentation should proceed (set points) and then after generating new information in the form of an algorithmic output, transmits information back to the fermentor, to a control unit capable of receiving information and translating that signal into a control action. Over and above this control function the operator can observe the flow of information (data), intervene if necessary (new information) and recall past information from a data storage system. Figure 9.5 shows an example of a fermentor control system architecture where the information exchange is indicated.

In Figure 9.5 there is flow of analogue and digital sensor data from the fermentor vessel to the control cabinet. The input data are marshalled in I/O (input/output) cards whose main function is to convert a mixture of signals to a common electrical signal type for the computer to interpret. These signals are passed to the Programmable Logic Controller (in this case) where control strategies (set points, etc.) are programmed. In the PLC the input data are fed into PID algorithms and other control functions, an output signal is fed back to the fermentor via I/O cards. The control signal returning to the fermentor may have to be transduced to a more usable format for the process plant, this may take the form of pneumatic signals scaled from the frequently used 4–20 mA to the equivalent 0–1 bar pressure. The PLC in Figure 9.5 is connected to the Operator Terminal which may be a personal computer with a VDU or a stand-alone plant terminal. If located on the process plant floor then the whole cabinet has to be splash and dust proof, often referred to as an IP55 rating.

The PLC is also connected to a Data Highway onto which other fermentor systems are connected. In Figure 9.5 all the

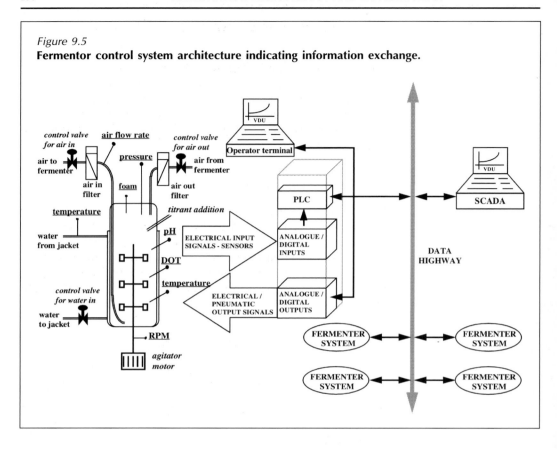

Figure 9.5
Fermentor control system architecture indicating information exchange.

fermentors are in communication with a Supervisory And Data Acquisition System (SCADA) whose function is to collect all the plant data and store it for later retrieval and possible interrogation. It should be understood that the SCADA is not a database, it generally cannot give answers to database type queries but using advanced graphics and trending graphs can present the operator with comprehensive information on plant status.

9.4.3 Fermentation plant safety

One of the over-riding functions of a fermentation control system is monitoring and action relevant to safety. A fermentor is a potentially dangerous piece of equipment using as it does live steam (to 135°C), over pressure (to 2 bar gauge +), acids and bases, as well as the microorganism and its products. The automated control system can ensure by program interlocks, etc. that the plant is operated with the maximum safety margins on every function, and specifying this is perhaps the most important part of the acquisition of a new control system.

9.5 Fermentor control specification

When planning for a new fermentor control system the specification for the system is crucial to the success of the project. The specification will be used to judge vendor's response and to subsequently quote and oversee the project to completion. Time and effort spent generating a comprehensive system specification is rewarded many times over during implementation and commissioning.

9.5.1 Specifying sequence control

One of the most important tasks when planning to automate a fermentor system, particularly for sterilization and other automated sequences, is to accurately specify what the sequences are to be automated and how this will be achieved. There are fixed unit operations associated with fermentor control and the specification will define these and indicate how this control will be achieved.

9.5.2 Fermentation unit operations

Taking a batch fermentation as a typical example, the unit operations to be selected might include:

- **Blank sterilization**: generally employed as a pre-batching cleansing sequence when all steam and drain valves are opened.
- **Medium batching**: the fermentor is usually in a safe state for opening and preparation. In the author's laboratory the safe state is defined by 'Standby'. In standby all valves are shut with the exception of valves venting the vessel interior to atmosphere. During batching probe calibration and insertion into the vessel must be complete before water is added to the vessel!
- **Medium sterilization**: the sterilization sequence will be partly dictated by the fermentor vessel configuration and geometry. Steam sterilization of the medium can be achieved either by direct steam injection or by indirect heating via a heat exchanger or vessel jacket. Whatever the mode of sterilization chosen it is crucial that the correct valves are operated in the correct sequence, not just for process integrity but also for safety.
- **Medium hold**: it is useful to maintain the fermentation medium in a safe state post-sterilization. It is at this time that final adjustments to the fermentor set-up can be made, e.g. pH adjustment via a titrant feed system or temperature alteration prior to inoculation.

- **Fermentor inoculation**: to introduce inoculum into the sterilized fermentor requires opening the vessel in a controlled manner and introducing inoculum using strict aseptic techniques throughout. If introducing the inoculum via peristaltic pumps then a reduction in vessel pressure will be required.

- **Incubation**: although not strictly a sequence there are clearly certain valves that must be opened to permit incubation of the culture to proceed under specified conditions. The principal function of the incubation sequence is to start the clock counting the hours elapsed since inoculation. During incubation the full functionality of the control system may be used to program the correct fermentation 'trajectory' for the run. This will involve specifying set points for all the main controllers including agitator speed, airflow, temperature and feeds if available as well as controls responding to more metabolic influences such as pH and dissolved oxygen.

- **Harvest**: It may be necessary to specify a sequence dealing with harvest operations. This could entail pre-chilling or suitable pH adjustment of the broth to permit easier product recovery operations. Harvest could also include 'killing' the vessel, i.e. sterilizing the vessel interior (with or without cells) prior to safe opening and cleaning.

- **Cleaning**: A number of options may be available here including full sterilization or heating in the presence of caustic detergents to fully automated Clean In Place (CIP) systems with complex valve operations of their own.

9.5.3   Vessel states

Fermentation unit operations may be termed vessel states and the transition between them must be strictly regulated in an automated plant, for example initiating automatic sterilization of the batch during the incubation should be avoided at all costs. State changes can be summarized in a state diagram (Figure 9.6).

Each one of the states in the diagram defines a part of the control program specifying sequences to be executed for effective control of the fermentation system. Navigating around the state diagram defines the safe operation of the vessel under automatic control. Certain transitions are under operator control such as from Standby to Medium Sterilization. Other states are attained automatically as a result of a particular sequence finishing such as the transition from sterilization of the Medium Sterilization to Broth Hold. There is within an automated control system the opportunity to initiate emergency action which could come about as a result of alarm settings or operator intervention. This is indicated in Figure 9.6 where a state called Emergency Hold may be attained from anywhere in the program, such a state can be programmed to do whatever is deemed appropriate for safe containment of the process, in the author's

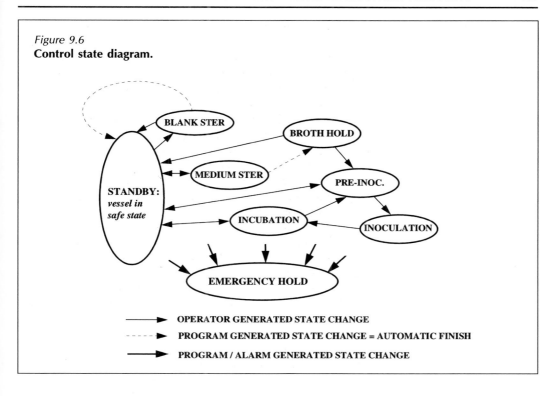

Figure 9.6
Control state diagram.

laboratory Emergency Hold constitutes a safe universal shut-down of the fermentation process sealing the vessel but retaining set points for possible restart of Incubation.

9.5.4 Sequence logic

Control of multiple valves on an automated fermentor requires the program to control the opening and closing of valves such that the state operation is effectively and safely completed. Each of the vessel states indicated previously and possibly many others have to be programmed taking best human operator practice and engineering constraints into account. Defining the sequence follows a pattern of working that ensures ambiguities are minimized and objectives are clearly stated. Taking one of the states in Figure 9.6, sterilization of the fermentor plus medium, the system developer must start with a complete description of the fermentor and the valves associated with sterilization (which is likely to be most of them). A comprehensive description will come from accurate drawings of the plant; these drawings are often referred to as piping and instrumentation drawings (P&ID, not to be confused with PID for control). The drawing will identify the units of control to be defined in the program. Modern operating systems will tend to function with structured code and it is possible to consider individual blocks of code controlling individual units of control.

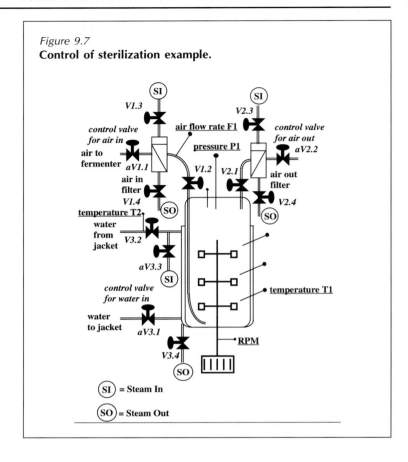

Figure 9.7
Control of sterilization example.

This can be illustrated with reference to Figure 9.7, a typical fermentor configuration (as a notional P&ID) is shown for sterilization, only those valves are shown that are relevant to this highly simplified representation.

The P&ID identifies the key elements for the following groups of valves:

- Air in group V1.X
- Air out group V2.X
- Jacket group V3.X

SEQUENCE LOGIC for *Sterilization operations* **(key words in BOLD)**

1. **START**: all valves closed
2. **Drain** jacket
3. **Heat**-up phase
4. **Direct** steam injection
5. Air **filter** sterilization
6. **Sterilization** temperature

Table 9.1 Valve descriptions for example fermentor (Figure 9.7)

Air in group		Air out group		Jacket group	
aV1.1	Analogue control valve: air to vessel	V2.1	Digital air out valve	aV3.1	Analogue water to jacket valve
V1.2	Digital Air ex: filter block valve	aV2.2	Analogue control valve for pressure	V3.2	Digital water from jacket valve
V1.3	Digital Steam to air in filter	V2.3	Digital Steam to air out filter	aV3.3	Analogue control valve steam to jacket
V1.4	Digital Steam out from air in filter	V2.4	Digital Steam out from air out filter	V3.4	Digital jacket drain/ condensate out valve

Table 9.2 Valve status chart for sterilization sequence logic

Valves	Valve status START	Drain	Heat	Filter	Direct	STER	Press	Cool	Ballast	Hold
Air in										
aV1.1	0	0	0	0	0	0	a = 1	a = 1	a	a
V1.2	0	0	0	0	1	1	0	0	1	1
V1.3	0	0	0	1	1	1	0	0	0	0
V1.4	0	0	0	1	0	0	$1 \Rightarrow 0$	0	0	0
Air out										
V2.1	0	0	1	1	1	1	1	1	1	1
aV2.2	0	0	a = 1	a	a	a	a	a	a	a
V2.3	0	0	0	0	0	1	0	0	0	0
V2.4	0	0	0	0	0	1	0	0	0	0
Jacket										
aV3.1	0	0	0	0	0	0	0	a = 1	a	a
V3.2	0	0	0	0	0	0	0	1	1	1
aV3.3	0	0	a = 1	a	a	a	a	0	0	0
V3.4	0	1	1	1	1	1	1	0	0	0

0: closed valve
1: open valve
a: analogue control 0 to 100% open
a = 1: analogue valve fully open
$1 \Rightarrow 0$: open to closed valve.

7. Air filter **pressurization**
8. Crash **cool**
9. **Ballast** air
10. Broth **hold**

9.5.5 *Flow charting*

P&ID drawings, valve descriptions (Table 9.1) and status charts (Table 9.2) are essential in defining the operation to be

Figure 9.8
Example flow chart for sterilization of a fermentor.

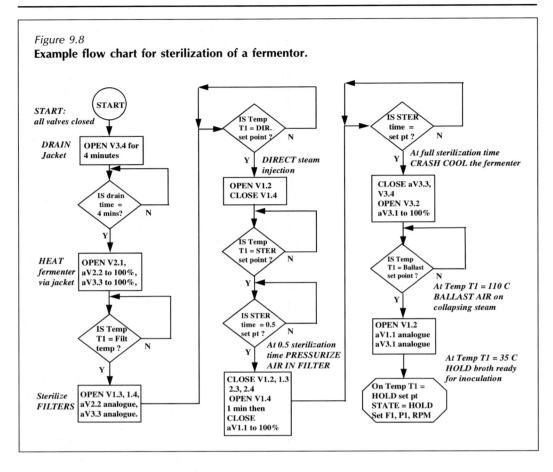

automated, however before code can be written the operation is translated into a flow chart of valve operation and decision gates that must be followed. The flow of code to accomplish the task will follow the chart (Figure 9.8).

The 'English' translation of the flow chart is approximately as follows:

- **START:** All valves are closed and remain closed unless instructed otherwise.
- **DRAIN jacket:** The jacket drain valves are opened for a specified time (dependent on jacket geometry).
- **HEAT:** Cycle commences by opening air out valves to allow for the escape of air and by opening steam valves to the jacket.
- **FILTER:** The sterilization of the air inlet filters must be complete before the medium so that sterile air can be admitted to break the vacuum caused by collapsing steam after sterilization

is complete. The commencement of filter sterilization may be dependent on the temperature of the medium in the fermentor.

- **DIRECT**: To assist in heating the medium to sterilization temperature, steam can be admitted via the air in filter and down the air sparge pipe. The point at which this occurs can be determined by the medium temperature.
- **PRESSURIZE**: At some point during the sterilization of the medium (0.5 total sterilization time?) the air in filters are pressurized to prevent ingress of contaminants prior to ballast air.
- **CRASH COOL**: When the sterilization time is complete the fermentor and its contents are 'crash cooled' (to minimize time at elevated temperatures) by shutting off the supply of steam and admitting cold water to the heat exchanger system.
- **BALLAST AIR**: Whilst the temperature of the medium is still above boiling the vessel is pressurized with sterile air to prevent the vessel pulling a vacuum and thus permitting possible ingress of contamination microorganisms.
- **BROTH HOLD**: When the temperature of the medium is at some predetermined level (35°C ?) the vessel is held in a quasi-incubation mode, ready for inoculation and/or other treatment.

In this simple example many different factors have been taken into account but the flow chart is far from complete, there is for example no mechanism in this for operators to enter set points. Again with this example there are only 12 valves, on a 4500 litre pilot scale fermentor used for research and production there may be 70+ valves organized in many functional groups, hence programming such a system will require many days/weeks of specifying, programming and testing before the fermentor control system may be commissioned.

9.6   Control of incubation

Specifying the sequence logic of a fermentor control system is only part of the control task, arguably the most important function of the fermentor control system is to control the incubation of the microorganism of interest. Returning to the case made for the control of fermentations at the start of the chapter, the purpose of the control system is to provide an optimal environment for growth and expression of an attribute associated with that growth. This is very difficult, one can pose the rhetorical question, what is the natural habitat for *Escherichia coli*? The answer given by many might be 'the intestinal tract of most vertebrate animals', but enteric bacteria can be isolated from many mesophilic environments and evolved probably billions of years before the advent of colons!

Therefore when faced with the prospect of defining the environmental conditions to maximize microbial productivity how does the fermentation scientist decide what conditions to apply? In answering this question a distinction should be made between fermentation development strategies for research and development organizations and manufacturing. Each function imposes different constraints. Within the manufacturing sector fermentation development will be investigating the factors associated with a well-established fermentation process where the objectives of a development fermentor will be to obtain relatively modest increases in productivity (leading to substantial financial savings on large-scale production) or to achieve reduction in 'cost of goods'. In a research and development environment the constraints usually come from working with a wide range of culture types, limited/no knowledge of those cultures, limited/no knowledge of the fermentation systems and with very short development times required.

9.6.1 *Specification for incubation control*

Specifying the control options for a fermentor system must then take into account not only the type of organisms to be grown but also what sort of control options will actually be required. The control options for a typical R&D fermentor system are described below.

9.6.1.1 *Temperature*

Control of incubation temperature will be achieved by the use of system of addition of heat and cooling. This control loop is fundamental to fermentor systems. The degree of accuracy of control during incubation will probably be in the order of a set point in the range 20°C to 50°C $\pm$ 0.1°C. Response times may also be important and controlled ramp rates of better than 1°C per minute may be necessary.

9.6.1.2 *Aeration*

For most industrially significant processes the microorganisms will be aerobic. Supply of air to the fermentor will need to be controlled typically in the range 0 to 2 VVM. The accuracy of the control system will need to be in the order of $\pm$1% of full scale deflection (FSD). This is generally a fast acting loop and obtaining adequate ramp rates is not a problem.

9.6.1.3 *Agitation*

The stirrer for a traditional fermentor is used to minimize gradients within the bulk broth, the larger the vessel the greater the potential for gradient (mass transfer) problems. Sufficient motor power has to be available to obtain uniform mixing times to be in the order of 1 to 2 minutes or less (determined from the addition of a marker into the bulk liquid and the time taken to homogeneity recorded). These fast mixing times may not be achievable in pilot scale vessels (3000 litres + working volumes) and these system constraints must be identified and resolved for example, in scale down experimental designs (modelling large-scale parameters in small-scale vessels). Agitation ramp rates may be high and the control system should be capable of ramp rates equivalent to 10% of FSD per minute.

9.6.1.4 *Pressure*

This control loop only applies to pressure-rated vessels. The control of pressure is required for accurate control of *in situ* steam sterilization as well as for some incubation regimes. The control requirement during incubation may come from maintaining adequate pressure for high oxygen transfer or to simulate hydrostatic head pressure in large-scale vessels in scale down experiments in smaller scale vessels. Control of pressure is critically linked to safety and safeguards within the control system must ensure that the vessel cannot be over-pressurized. Pressure can be a very fast-acting loop and obtaining adequate ramp rates is not a problem. The function of this loop may be in conflict with the aeration loop and the fermentation scientist has to bear in mind that control of these two loops may be incompatible. For example, a set point for pressure near the safe operating limit of the vessel may be impossible to achieve with extremely low air flow rates (0.1 VVM).

9.6.1.5 *pH*

Most industrial fermentations need to be run within a certain range of pH values for maximal productivity. Although media are formulated in such a way that ensures a certain level of buffering capacity 'built in', there is often a requirement to control the pH away from this 'natural' value. This is achieved most often by the addition of acid or base titrants. Addition of acid or base titrants requires a feed system of some description, under these circumstances the control of the feed system is subordinated to the pH controller, in other circumstances the feed system may act under the direction of another controller or independently; this will be discussed in more detail below. When tuning a pH control loop it is very difficult to define the 'correct' PID

Figure 9.9
pH control by titrant addition.

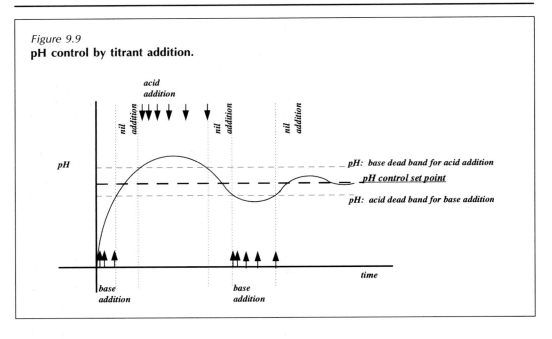

settings because the strength of the titrants will vary therefore the 'gain' of the controller (i.e. how quickly a response will be achieved) will also vary greatly. This problem of adequate pH control can be addressed by user definable 'dead bands' around the set point where no control action is initiated. The purpose of the dead band is to allow for the 'natural' buffering capacity of the medium to have sufficient time to act so that titrants are not added close to the set point on a detuned loop which will be the case with variable titrant concentrations.

In Figure 9.9 the response of a pH control loop following acid or base addition is shown. It will be noted that there is a proportional action on the additions of titrant made, i.e. the larger the 'error' the more frequent the control action. The closer the measured pH gets to the set point the longer the delay between each addition of titrant, however when the dead band threshold is crossed all titrant addition stops and the natural buffering capacity of the medium results in an approach to the set point within the dead band limits. The setting of the dead band limits may be asymmetrical about the set point depending on the criticality or tendency of the fermentation to be acid or base tolerant. Mineral acid titrants may be substituted for organic acids or the principal carbohydrate energy source such as glucose. In this case the buffering capacity or delay equates to the metabolic rate for the consumption of the sugar and protons generated by catabolism will require a delay of some minutes before a noticeable effect on the pH will be detected. This type of control of pH by metabolic action is really a means of permitting the microorganism to autoregulate the supply of carbon.

9.6.1.6   Dissolved oxygen

The control of dissolved oxygen in aerobic fermentations may be critical to the successful outcome of that fermentation. The requirement for oxygen may be very high during the rapid growth phase of a batch culture and oxygen limitation may result in inadequate growth and incomplete oxidation of the primary energy source. The control of dissolved oxygen is generally achieved by increasing the 'driving force' for the mass transfer of oxygen from the gaseous phase to the liquid phase. The methods or outputs available to the control system to achieve this include increasing the agitator speed which generally increases the shear on the air bubbles making them smaller and increasing their surface area available for gaseous exchange.

The second output that may be programmed is to simply increase the air flow rate, ensuring a greater volumetric airflow through the liquid. The last output that may be incorporated into a dissolved oxygen controller is vessel overpressure which will have the effect of increasing the hydrostatic head pressure and preventing rapid flushing of 'precious' air from the fermentor.

All these outputs can be specified for a dissolved oxygen controller either individually or in any combination of all three. The output of the dissolved oxygen controller (master) must be cascaded onto the closed loop controllers for agitator, airflow and pressure (slaves) as required. The consequence of this is that the function of the closed loop controller will be subordinated to that of the dissolved oxygen controller to achieve the desired dissolved oxygen tension in the fermentation broth.

This type of control may be considered a replenishment system where the microorganisms' metabolism is the oxygen sink and the outputs of the controller replenish the resulting deficiency. In setting up a dissolved oxygen controller it may be necessary to set default set points for each of the closed loop controllers such that at the start of the fermentation a baseline level of agitator speed, airflow rate and vessel pressure will permit adequate mass transfer of oxygen for microbial metabolic action to ensue. As the microbe respires the depletion of dissolved oxygen will occur and the output of the dissolved oxygen controller will increase to compensate. Depending how the DOT control loop is specified the outputs can be scaled to one or more of the specified closed loop controllers.

In Figure 9.10 the dissolved oxygen controller (master) is linked to all three closed loop controllers (slaves). As the DOT falls the oxygen controller output will rise and this will cause an increase in agitator speed from a default closed loop set point to a maximum rpm specified for the DOT output. In the example the DOT continues to fall as the output of the controller passes a threshold equivalent to one-third the total output range of the DOT controller. At this point the second specified closed loop

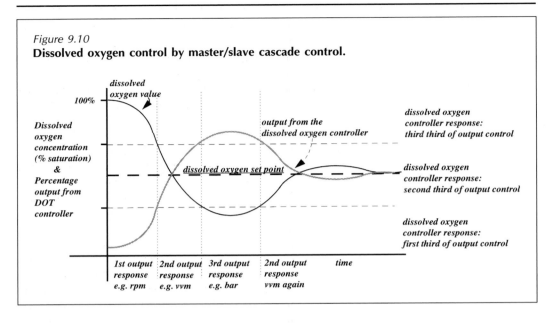

Figure 9.10
Dissolved oxygen control by master/slave cascade control.

output is subordinated to the DOT controller and takes over from agitator speed leaving the set point for the first output at its specified maximum. The second output, in this case airflow, increases to try and compensate for the fall in dissolved oxygen and when this output limit is reached (second third of the DOT controller output) then the third closed loop (in the example vessel pressure) takes over, leaving airflow at a maximum value. Eventually the rate of replenishment equals the rate of oxygen depletion and DOT is controlled in the example with rpm at its maximum set point, vessel pressure at its minimum and airflow responding as the principal output of the DOT controller as a PID loop.

As well as control by oxygen replenishment, it is also possible to initiate dissolved oxygen control by 'depletion'. With the slave outputs set to relatively high output values, medium batched with little or no energy source present can have the dissolved oxygen tension controlled by the regulated feed of, say, carbohydrate. Under these circumstances the feed system becomes the cascaded slave output and only delivers glucose at a rate equivalent to the metabolic rate of the organism under fixed oxygen mass transfer limitations set by the closed loop outputs (speed, air and pressure).

9.6.1.7   Feed systems and antifoam

Most fermentor systems will be equipped with a mechanism for introducing various feed solutions under aseptic conditions. The method for achieving this will depend largely on the size and configuration of the fermentor itself. On smaller laboratory scale vessels, feeds will be controlled by the use of peristaltic pumps working on flexible walled tubing. On larger vessels such as those found in the pilot plant (> 10 to 20 litres) it is possible that purpose designed and constructed addition vessels independently sterilizable will deliver feeds using shot-wise additions of feed solutions.

Whatever the method of addition chosen it is very likely that feeds will be pulsed in some way and therefore the variables available for control are pulse width (i.e. size of liquid 'shot') and interval between pulses. In the author's laboratory it has been found that fixed pulse width or liquid volume shot size and variable intervals provide a satisfactory control option for the addition of feeds. The set point for feed controller can be a number of shots in a specified interval of control or the feed controller can be cascaded onto a master controller such as pH or dissolved oxygen. The output of the feed controller can be linked to the interval such that the larger the output of the controller (further away from the desired number of shots) the shorter the interval and hence the more rapid the feed rate. Calibration of a feed system is difficult and will be affected by tubing age and vessel backpressure. It is possible to introduce another sensor to take over the control of feed additions, i.e. a balance for the feed reservoir, this type of control loop for feeds can provide very accurate and non-pulsed control of feed flows but costs may be high and it is only really applicable to smaller scale vessels (< 20 litres).

Antifoam addition is variation of the feed control loop where the sensor is a contact probe detecting rising foam. The control variables here may be time related. To prevent overaddition of antifoam agent (which may destroy dissolved oxygen control by bursting air bubbles) a splash time interval can be specified whereby the contact probe would need to be covered for more than a few seconds to initiate antifoam addition. Similarly if the probe remains covered following antifoam pumps being switched on then secondary control action can be specified whereby the airflow rate and/or agitation speed may be reduced to minimize the generation of foam. Clearly if this happens then the dissolved oxygen controller will be affected but the output of the antifoam controller may be the ultimate master controller of closed loops to prevent loss of broth by foam-out.

9.7   Advanced incubation control

The control options specified above represent the basic control elements for most fermentation development purposes. However, many fermentation development programmes will require more than just a system of fixed set point control and that is where we can return to some of the observations made at the start of the chapter. To close this chapter there follows a description of an advanced Programmable Logic Controller in the author's laboratory which is capable of more advanced fermentation control options.

The wild type microorganism in its natural habitat (whatever that is) as previously discussed, is subject to transient environments. When a fermentation scientist attempts to grow the microbe and 'encourage' it to express the desired attribute, a fermentation regime will be imposed that will only represent a tiny fraction of all the influences that the organism will encounter. Given that most fermentations will be run under fixed set point regimes, the environment that is thus imposed will effectively be a 'snapshot' of the full range of conditions that have brought about the expression, through evolutionary pressure, of the desired phenotype. Another option for control therefore is to provide a mechanism whereby events beyond the fixed set point control regime can be specified.

9.7.1   Fermentation profiles

A typical fermentation control profile is shown below (Figure 9.11). The main features of the profile are captured in Table 9.3 below. When analysing the progress of fermentation it is sometimes useful to identify the key features and in which phase of the growth of the organism they occur. For this purpose it is useful to use standard definitions for the growth phases although their strict interpretation is obviously open to debate. In the example given, the following features become apparent:

Figure 9.11

Typical fermentation profile for a filamentous microorganism producing a secondary metabolite.

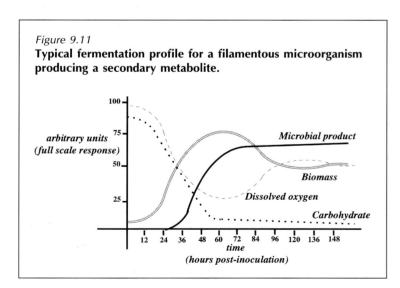

Table 9.3 Fermentation profile data for key analytes at various growth phases ('snapshots')

Analyte	Time (h)	Value (arbitrary)	Growth phase
Biomass	12	10	lag
	24	25	acceleration
	36	50	exponential
	48	70	deceleration
	60	75	stationary
	84	60	decline
	148	50	harvest
Carbohydrate	0	85	inoculation
	24	70	acceleration
	36	35	exponential
	48	20	deceleration
	60	10	stationary
	148	5	harvest
Dissolved oxygen	0	100%	inoculation
	24	80%	acceleration
	36	40%	exponential
	48	30%	deceleration
	60	25%	stationary
Microbial product	24	0	acceleration
	36	10	exponential
	48	35	deceleration
	60	50	stationary
	72	60	decline
	148	65	harvest

- Harvest time is a long time after product accretion has ceased (this is frequently the case with old processes that have been 'inherited').
- Carbohydrate uptake is largely linear over the growth phase of the organism, and is depleted by 60 hours.
- Biomass accretion occurs between 0 and 60 hours and then following substrate depletion goes into the decline phase.
- Dissolved oxygen profile is the mirror image of the growth profile.
- Product accretion has its onset at 24 hours and is complete by 72 hours.
- Harvest time using this protocol can be at 72 to 84 hours.
- Dissolved oxygen is not limiting, therefore under this fixed set point regime would higher biomass yield more product (increasing fermentor vessel volumetric productivity)?
- Carbohydrate is depleted by 60 hours, a further carbohydrate feed at either this point or when the DOT was less than 40% of saturation may promote a further product accretion phase.

- pH is not shown here but the level is likely to fall between 12 and 60 hours and this may require pH control by titrant addition.

This type of profile is typical of the kind that would be obtained with fixed set point control for the principal closed loop controllers (not including DOT control).

9.7.2   Event-tracking control

As indicated above with the carbohydrate feed option it is possible to identify an event in the fermentation which serves to trigger another control action. Hence an event is a specific change in state or time or any combination of changes that can initiate a new event. In the case cited a carbohydrate feed could be initiated at 60 hours post-inoculation (time-based event) or if the dissolved oxygen measurement fell below 40% of saturation (analogue value event). Computer control systems, particularly those specified by the customer, can easily accommodate program decision gates which will test the status of the fermentation and apply another control action on the fermentation if the decision gate criteria are met. To define what the decision gates should be, and what values of analyte or combinations of analyte values should be used requires the fermentation scientist to establish set points for control loops that he or she can control and then observe the effect on analytes that do not have on-line sensors (biomass, carbohydrate and product in this example).

As soon as the fermentation scientist wishes to use event-based control then further options may be sought to extend this capability. In the author's laboratory the PLC fermentor control system has been programmed with four types of user definable events:

- Time-based events: become true at a specified number of hours post-inoculation.
- Analogue value events: become true when a process value or combination of two process values exceed a threshold limit.
- Elapsed time events: become true at a defined time after another event has occurred.
- Boolean events: logical combinations of any two other events using standard Boolean operators.

The trigger events for the most part can be almost any process signal or event 'flag' indicating the status of that event, in addition, system events or alarm levels can be specified (this apart from warning alarms is a very powerful use of alarms signals). The events can be organized in any combination giving virtually limitless fermentation control strategies. The events themselves can initiate new set points, ramp rates (rates of change between

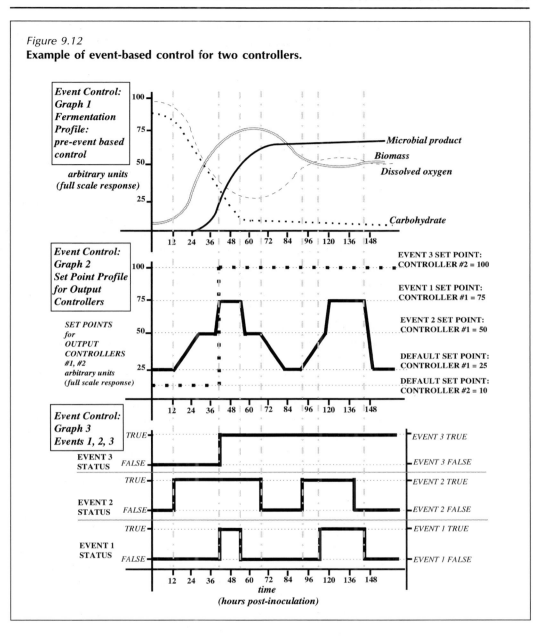

Figure 9.12

Example of event-based control for two controllers.

set points) or new events. It is possible to 'latch' initiating events such that dependent events will remain true even if the initiating event is no longer true.

In Figure 9.12 three graphs are shown for event-based control. Graph 3 shows the status (true/false) of three fermentation process events. Graph 2 shows set points for two output controllers (these could be any type of controller as discussed previously) to be driven by the specified events (1, 2 and 3). Graph 1 is the fermentation profile against which the new set point control regime is to be imposed. A summary of the main changes is given below:

- Event 2 becomes true first (e.g. time value = 12 hours) and the set point for controller #1 is changed from the default set point (arbitrary value 25) to the event 2 set point (50) – control action initiated at end of lag phase.
- The rate of change from one set point to the other can also be specified and in the graph it is comparatively slow – ramp rate control action over acceleration phase.
- At approximately 40 hours event 1 becomes true. The set point of controller #1 changes from event set point 2 to event set point 1 (arbitrary value 75) with a faster ramp rate – control action at end of the exponential phase, start of deceleration phase.
- Event 1 and 2 are both true at this time. This initiates event 3 also becoming true, the set point for controller #2 changes from arbitrary value 10 to 100 (could be a feed rate for example) – example feed rate response could be in response to deceleration phase.
- At approximately 50 hours event 1 then no longer is true and set point controller #1 returns to event set point 2 level (50). Note event 3 set point is 'latched' and remains true for the remainder of the batch – start of stationary phase.
- At approximately 70 hours both event 1 and 2 are no longer true and the default set point for controller #1 returns to the default level of 25 (approximately 84 hours) – 70 hours end of product accretion phase.
- At approximately 90 hours event 2 becomes true again, and the event 2 set point (50) for controller #1 is used. Before the event 2 set point is reached, event 1 becomes true and the event 1 set point (75) is used again (faster ramp rate). Event 1 set point is in use for controller #1 by 120 hours – recovery from decline/death phase by 90 hours (scavenged substrates?).
- At approximately 140 hours event 2 is no longer true, and a short time later event 1 is no longer true, at this point the controller #1 set point is reduced from 100 to 25 (default set point) – harvest time.

The results of this event-based control will be judged by comparing the fermentation profile (as shown in Figure 9.12: Graph 1) with the one generated for the event control regime.

9.7.3 *Boolean control and rule generation*

It may be seen from the example in Figure 9.12 that complex patterns of control can be imposed on the fermentation and

Table 9.4 Boolean truth table for fermentation control

Boolean operator	Event 'a' (pH > 6.5) status	Event 'b' (time < 24 h) status	Comments (example event combinations)
AND	Yes	Yes	BOTH: analogue and time i.e. pH > 6.5 AND < 24 h true
OR	Yes	No	EITHER or BOTH:
	No	Yes	i.e. pH > 6.5 true OR time < 24 h true OR
	Yes	Yes	both true
XOR	Yes	No	EITHER true (but *not* both):
	No	Yes	either pH > 6.5 true OR time < 24 h true
NAND	No	No	EITHER or NEITHER (but *not* both):
	No	Yes	i.e. neither pH > 6.5 OR time < 24 h, OR pH,
	Yes	No	OR time, but NOT pH AND time true
NOR	No	No	NEITHER: neither pH > 6.5 OR time < 24 h true

more or less significant changes to the growth environment and expression of phenotype will follow. It is however difficult for the fermentation scientist to know which fermentation control regime to impose. This lack of knowledge comes from not knowing how the organism will respond to changes in individual control loops but also what the interactions with other control loops will be. To address this, Boolean logic can be employed to introduce an element of control by 'choice'. With Boolean logic it is possible to present options or 'choices' from which the control system can select a pre-programmed path for control. This type of experiment is then generating a new kind of response variable, one based on the control path selected by the metabolism or response of the total fermentations system, where:

Fermentation system = stainless steel vessel + valves + sensors + services (air, electricity, etc.) + medium + microorganism + microbial metabolism + expressed phenotype

The five Boolean operators that can be used are AND, OR, XOR, NAND and NOR. On their own these terms are rather impenetrable. Table 9.4 summarizes what they mean, examples given all represent truth statements affecting just two fermentation events – a pH value and a time value.

If Boolean options or choices are presented to the system then the path selected by the total system can represent a 'Rule' by which the response of the microorganism to an imposed environment can be described. To illustrate the principle of control by Rules, Figure 9.13 shows a simple XOR Rule statement.

The Rule defined in Figure 9.13 is remarkably simple and helps to define a glucose feed regime by pH and dissolved

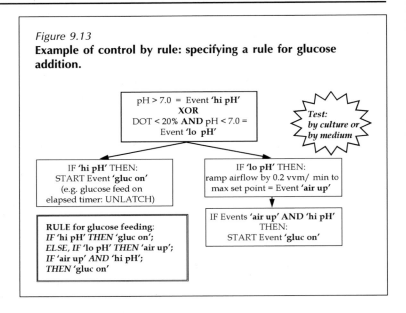

Figure 9.13

Example of control by rule: specifying a rule for glucose addition.

oxygen control. Having established this Rule a fermentation experiment can then test which route is 'chosen' under defined conditions of medium or mass transfer for oxygen and so on. Another powerful use for control by Rules comes with culture or mutant evaluation where the 'chosen' route may indicate a propensity towards one pattern of metabolism or another, which may have been induced within a putative mutant culture. Clearly this is just one of many Rules that could be introduced and libraries of Rules or elements that make up Rules can be constructed to provide an array of control options to explore microbial metabolism under controlled conditions.

9.7.4 *Summary of event and non-stable set-point control*

A summation of the possible interactions for event-based control is shown in Figure 9.14, here control options are defined as possible inputs and outputs to an experimental system where events become control loops themselves evaluating whether conditions defined by Rules are satisfied. This type of control system is an extension of the standard media and control recipes normally prepared and fermentation response data will produce a data set of on-line, off-line and derived data which generate Rules for high productivity.

Returning to some of the opening discussion points in this chapter, the fermentation scientist is being asked to establish the 'correct' conditions for a microorganism to express a desirable attribute or phenotype in a totally artificial environment; event

Figure 9.14
Control of fermentation by event loops and set-point rules.

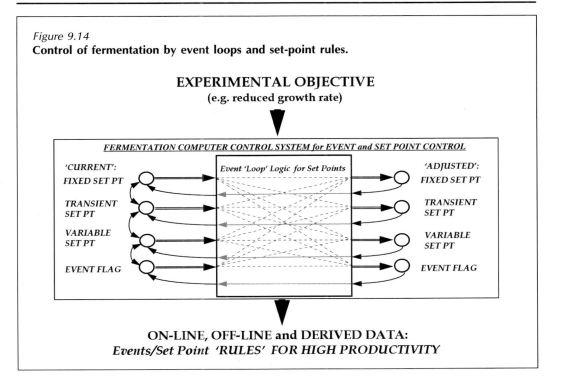

EXPERIMENTAL OBJECTIVE
(e.g. reduced growth rate)

FERMENTATION COMPUTER CONTROL SYSTEM for EVENT and SET POINT CONTROL

'CURRENT': FIXED SET PT

TRANSIENT SET PT

VARIABLE SET PT

EVENT FLAG

Event 'Loop' Logic for Set Points

'ADJUSTED': FIXED SET PT

TRANSIENT SET PT

VARIABLE SET PT

EVENT FLAG

ON-LINE, OFF-LINE and DERIVED DATA:
Events/Set Point 'RULES' FOR HIGH PRODUCTIVITY

and Rule based control together with transient and variable set point control (non-stable) may help establish what unique set of factors and their interactions support the expression of a rare and valuable phenotype.

9.8 Other advanced fermentation control options

This chapter has focused on a tiny part of control technology available for fermentation systems. The reader should be aware that other control philosophies are in use and this technology is certain to continue to change and develop. Among some of the control philosophies currently in use or in advanced development are described below.

9.8.1 *Knowledge-based systems (KBS)*

These systems can be considered an extension of the sort of control capability described previously where previously held 'expert knowledge' is captured in a control system often referred to as an expert system. With this type of control already established facts and data associated with a process can be held in a data base of knowledge against which future decisions can be made. The principal difference with expert systems and the event-based control described here is that knowledge-based systems will only work on existing knowledge and won't necessarily generate new knowledge. The power of expert systems lies in their ability to execute potentially hundreds of rules per second controlling many functions associated with plant operation simultaneously. Early warning patterns can be recognized from this previous knowledge to alert either human or computer control operations of an excursion from normal operating parameters.

9.8.2   *Artificial neural networks (ANN)*

Much discussion has been made in recent years of 'intelligent' computer systems. By intelligent systems we mean systems capable of learning. The mechanism by which this occurs is beyond the scope of this chapter but for the most part is very similar to the control loop principle whereby an input to a system is modified by attempting to minimize the error between desired and actual outcome. The use of neural networks in fermentation will come from the ability of these systems to recognize patterns and take actions on unknown data based on those learnt patterns. It is possible that the event-based system described here will be augmented by neural net ideas to help with obtaining data or process inferences not available by direct measurement. ANNs do not in themselves assist in providing data on the exact nature of process input interactions because the training and optimization of an ANN is a pure black box model of the process.

9.8.3   *Genetic algorithms (GA)*

In this case control philosophy has been influenced by modern biological genetics. Again it is not possible to describe how these systems work here but the principle that should be understood is that GAs are capable of searching vast areas of experimental space and by a process of 'natural selection' of rules or algorithms that come closest to minimizing some process 'cost' function tend to optimize on a process goal. This type of system may assist in rule induction by eliminating rules not 'fit' for the process environment.

9.8.4   *Modelling*

For many years there has been a disparity between what modelling could apparently offer and the acceptance and use of modelling by the process community for everyday control and optimization problems. This may now be changing with the advent of advanced user-friendly software to assist in forming and testing model predictions. There are many kinds of models that may be constructed, in attempting to describe a complex system the key to approximation is simple models, i.e. relating carbon dioxide evolution to biomass or growth estimations. These types of relationships can be defined relatively easily by curve fitting raw data and then predicting on the basis of alteration of parameters defined in function generated from curve fits. When several functions describing microbial growth and productivity have been created then they can be combined in a mathematical relationship potentially useful in predicting fermentation outcomes. If models are successful then thousands of fermentations can be run in software before committing to expensive stainless steel vessels.

Summary

This chapter has attempted to elucidate the complexity of fermentation systems and the difficult task facing the fermentation scientist who has to elucidate the key features of a fermentation to significantly enhance microbial productivity with a minimum expenditure of resource. Fermentation development has traditionally used an heuristic approach to refine empirical knowledge. The requirement for empiricism is still very much present but modern fermentation control systems and techniques should allow for a more systematic approach to fermentation optimization (see Chapter 5 for more details). It is likely that modern computing techniques such as event-tracking and rule-based controls will augment the application of microbial physiology to solving problems of applied microbiology. Microbial physiology knowledge itself in turn may benefit from advanced control ideas by increasing understanding how a microorganism can successfully respond to its environment.

Suggested reading

Bailey, J.E. and Ollis, D.F. (1986) Instrumentation and control, in *Biochemical Engineering Fundamentals*, 2nd edn, Maidenhead: McGraw-Hill, pp. 658–722.

Beluhan, D., Gosak, D., Pavlovic, N. and Vampola, M. (1995) Biomass estimation and optimal control of the fermentation process, *Comp. Chem. Eng.*, **19**(Suppl.), 387–392.

Blackmore, R.S., Blome, J.S. and Neway, J.O. (1996) A complete computer monitoring and control system using commercially available, configurable software for laboratory and pilot plant *Escherichia coli* fermentations, *J. Ind. Microbiol.*, **16**, 383–389.

Chen, W., Graham, C. and Ciccarelli, R.B. (1997) Automated fed-batch fermentation with feed-back controls based on dissolved oxygen (DO) and pH for production of DNA vaccines, *J. Ind. Microbiol. Biotechnol.*, **18**, 43–48.

Diaz, C., Dieu, P., Feuillerat, C., Lelong, P. and Salome, M. (1995) Adaptive control of dissolved oxygen concentration in a laboratory-scale bioreactor, *J. Biotechnol.*, **43**, 21–32.

Gregory, M.E., Keay, P.J., Dean, P., Bulmer, M. and Thornhill, N.F. (1994) A visual programming environment for bioprocess control, *J. Biotechnol.*, **33**, 233–241.

Kong, D.Y., Gentz, R. and Zhang, J.L. (1998) Development of a versatile computer integrated control system for bioprocess controls, *Cytotechnology*, **26**, 227–236.

Kurtanjek, Z. (1994) Modelling and control by artificial neural networks in biotechnology, *Comp. Chem. Eng.*, **18**(Suppl.) 627–631.

Omstead, D.R. (ed.) (1990) *Computer Control of Fermentation Processes*, Boca Raton, USA, CRC Press.

Onken, U. and Weiland, P. (1985) Control and optimisation, in Rehm, H.-J. and Reed, G. (eds) *Biotechnology*, Vol. 2, Weinheim: VCR Verlag, pp. 787–806.

Romeu, F.J. (1995) Development of biotechnology control systems, *ISA Trans.*, **34**, 3–19.

Sys, J., Prell, A. and Havlik, I. (1993) Application of the distributed control system in fermentation experiments, *Folia Microbiol.*, **38**, 235–241.

Wang, H.Y. (1986) Bioinstrumentation and computer control of fermentation processes, in Demain, A.L. and Solomon, N.A. (eds) *Manual of Industrial Microbiology and Biotechnology*, Washington DC: ASM, pp. 308–320.

10 Command Control in the Fermentation Industry

H. Blachere, B. Dahhou, G. Goma, G. Roux, J.-P. Steyer, E. Latrille and G. Corrieu

10.1 Needs and limitations

The establishment of reliable processes with increased efficiency and reduced operating costs is of primary importance in the fermentation industry. Unlike strain development programmes, through mutagenesis and genetic manipulations (see Chapter 5 for more details), which are carried out with the sole view of increasing carbon flux to product formation, command control systems are concerned with improving the reliability and reproducibility of the overall process.

The application of command control systems is limited to a few simple but fairly robust PID controls in which closed loops or feed-forward control strategies are often used. These control systems rely on a combination of on-line and off-line measurements using automatic or manual sampling without compromising the process or increasing the risk of contamination. Such strategies have enabled considerable optimization of secondary metabolite fermentations, in particular in antibiotic production. The use of specific probes, however, remains very limited mainly because of their instability following sterilization on the one hand and their susceptibility to fouling on the other.

The high cost associated with many fermentation processes makes optimization of bioreactor performance through command control systems very desirable. Since the majority of fermentation processes are either batch or fed-batch, it is important that we minimize the 'turn-around' time between different cycles to maximize output and, in turn, productivity. Such an objective can be achieved to a large extent through automation of certain aspects such as medium preparation, sterilization, vessel filling and emptying, and so on.

Clearly, control of fermentation is recognized as a vital component in the operation and successful production of many industries. The guidelines of good manufacturing practice (GMP) with respect to command control systems stipulate the following requirements:

- A complete description of the methodologies used for its development, installation and operation
- Validation of the system with respect to hardware, operation and applications
- Documentation of specific validable activities, e.g. performance and staff responsible
- Evidence confirming that each element of the hardware and software perform its function reliably and in accordance with documented specifications.

Naturally, companies which satisfactorily fulfil the above criteria enjoy a distinct advantage over their competitors. The recent upsurge in the developments of command control systems for the monitoring, automation and control of fermentation processes is largely due to the development of a new generation of computers that are low cost yet very powerful.

10.1.1 *Recent trends in bioreactor control*

In the last decade, manufacturers of bioreactors have provided proprietary software able to produce set-point profiles, reporting and producing graphs of many different parameters. Most of these software programs are based on MS-Windows. More recently, however, a number of different manufacturers have generated a whole host of different control programmes that are both versatile and capable of performing the most difficult of tasks. For example, the following companies offer software programs with a wide range of applications:

- APPLIKON offers a wide choice of software with sophisticated interpreters for the writing of equations and the calculation of virtual measurements at the laboratory, pilot plant and industrial scale
- B. BRAUN BIOTECH offers MFCS/Win, which is capable of managing as many as 16 bioreactors as well as configuring set-point profiles for any variable. It is also capable of controlling the dilution rate, calculating the growth rate, interfacing with other devices such as analysers and archiving of data for validation
- INCELTECH with INCELSOFT offers a software able to detect automatically the configuration of 128 controllers dispatched on a system of 10 fermentors
- LSL BIOLAFITTE proposes MENTOR, a software able to control large industrial systems
- NEW BRUNSWICK SCIENTIFIC proposes AFS BioCommand, managing 8 bioreactors, which is able to make

data logging, compute derived process values, altering set-points without writing equations, automatic feeding and data archiving for FDA validation.

These proprietary software programs are generally written in Basic, Pascal or C languages and as such are very difficult to modify to satisfy a particular need. To overcome this difficulty, bioreactor manufacturers have agreed to adopt the 'Microsoft Windows' as a general platform for the manufacturing of software programs with Window 95 for programs intended to be used at the laboratory scale and 'NT' for large-scale processes. The use of this multitask platform facilitates dynamic exchange of data between external programs and real time control systems, thus bringing about a new generation of 'intelligent' process control.

The use of software programs in the supervision and management of fermentation processes has also undergone dramatic developments to satisfy the market demands which are apparently very large. A number of companies have invested very heavily in the development of a new range of high performance, reliable and flexible programs that fulfil and satisfy all the validation criteria required for successful validation. Four manufacturers have already used this new approach:

- ADAPTIVE BIOSYSTEMS offers the software package 'BioView98', based on LABVEW, which is capable of controlling up to 8 fermentors. BioView98's strong features include adaptability to different fermentors, profiling, sequencing/recipes, off-line data, ancillary equipment and flexible interfacing
- INFORS SOFTWARE (IRIS NT4) for data logging and control provides great flexibility in terms of what equipment can be linked to the system as well as automation without the need to write control language sequences
- LSL BIOLAFITTE introduced in 1997 SYMPHONY, based on LABVEW also for control of laboratory bioreactors with a PLC card inside a PC
- INCELTECH is the first bioreactor manufacturer to use fully this new approach with standard Siemens PLC and BOSS, an application based on INTELLUTION's FIX software. The large number of FIX applications (more than 30,000, some of them with validation) give to the end-user a real insurance for maintenance and validability of the software.

BOSS, illustrated in Figure 10.1, is the first of a new generation of command-control systems for laboratory, pilot and industrial systems, offering:

- Standard PLC controllers with field Modbus for bioreactors networking

Figure 10.1
Bioreactor Operating and Supervision System (BOSS).

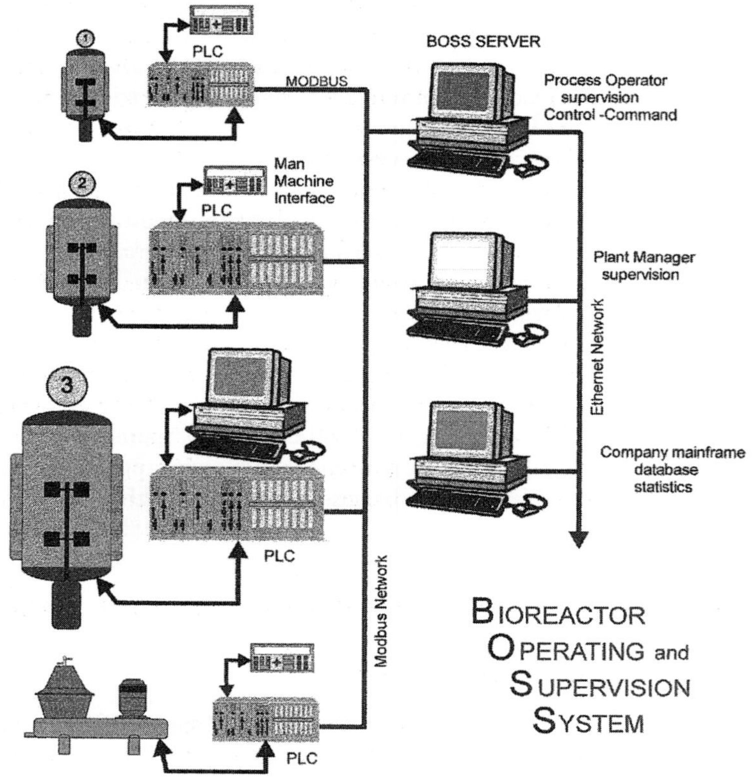

- Microsoft Windows platform versions 95 or NT
- Standard Ethernet external bus for connection to mainframe computer
- Code accessible to the user for maintenance or modification
- Locked files for validation purpose

This new generation of software satisfies the needs of both research and industry and conclusively resolves the problems outlined in the previous paragraphs:

- PLC allows unsurpassed reliability, modularity and expandability of control loops independently of the supervisor
- Power and flexibility of the supervisor offers all functionality for sequence programming, data logging and reduction, calculations, set-point profiles, recipe, automatic reporting, teleprocessing, etc.

- Internal and external networking allows interfacing with sophisticated on/off-line analysers like autoanalysers, mass-spectrometers or chromatographs.

The most significant progress thus far has been the ability to secure interface between the external programs, running in MS-Windows, and the real time program. This, in turn, allows the use of other capabilities such as fuzzy logic, neural networks or expert systems for the detection of failures or anomalies in the behaviour of the system.

Indirect measurements form much of the basis of estimation techniques. Adaptive control, fuzzy logic and expert systems have been intensively tested at laboratory level. Some industrial applications have been successfully implemented. Several different control approaches were developed over the years for model-based control strategies and adaptive control (Bastin *et al.*, 1990; Shimizu, 1993). Recently, a number of different control approaches have been developed, e.g. adaptive control (Bastin *et al.*, 1990; Shimizu, 1993), thus offering more scope for effective control of fermentation processes. The potential and applicability of these approaches are described below.

10.2   The adaptive control approach

10.2.1   *Introduction*

There are two main reasons for using adaptive approaches. First, because the dynamics of biological processes are both non-linear and non-stationary in nature; secondly, because classical controllers have proved inadequate in describing the overall behaviour of biological processes. Any problems that may rise whilst adaptive control problems are in operation can be resolved using either a linear approach based on input/output formulation (Dahhou *et al.*, 1993) or a non-linear approach based on the direct exploitation of the non-linear structure of the model in operation.

In the following sections, we shall describe and, in turn, discuss experimental results obtained under 'mono' and 'multivariable' adaptive linear control systems as well as under 'mono' and 'multivariable' adaptive non-linear control systems. This latter set of adaptive controllers was designed taking into account the non-linear and non-stationary nature of biological processes. One of the principal objectives of adaptive control systems is to compensate for any variations caused by the non-stationary or/and non-linear nature of the system under investigation. The following structure will be used for the discussion of our results: in section 10.2.2 the experimental design is

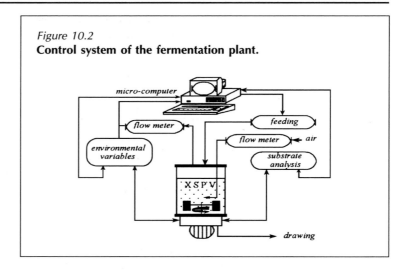

Figure 10.2
Control system of the fermentation plant.

presented; in section 10.2.3 the process model is described and the different control objectives are given; in section 10.2.4 the four control algorithms used are briefly formulated; and finally in section 10.2.5 the efficiency of the proposed algorithms is compared and discussed using the data obtained during the course of this study.

10.2.2   Experimental design

A schematic diagram of the experimental design representing alcohol fermentation is given in Figure 10.2 in a simplified form. This is different from batch fermentation where feeding and withdrawing are both suppressed.

10.2.2.1   Strain and experimental conditions

The strain used in our studies was *Saccharomyces cerevisiae* UG5 and the medium used was glucose minimal medium, pH 3.8. For batch and continuous processes, a two-litre capacity fermentor equipped with magnetic agitator, temperature and pH controls was used. During the course of continuous fermentation, a level-sensor was used to maintain constant volume within the bioreactor. Fresh medium and substrate were supplied to the reactor at a specific rate, computer controlled, and as such the dilution rate could be varied to regulate the concentration of the substrate in the reactor. For the fed-batch fermentation, a bioreactor with a 20-litre capacity was used. During fermentation, the stirring speed, the temperature and the pH were all controlled and continuously monitored. The operating variables

Table 10.1 List of operating variables and parameters for batch, continuous and fed-batch fermentations

	Batch	Continuous	Fed-batch
Active volume (l)	1.34	1.34	6–16
Temperature (°C)	30	30	30
Stirrer speed (rpm)	200	200	300
pH	3.8	3.8	3.8
Aeration (l h⁻¹)	3	3	32
Feed flow rate (l h⁻¹)		0–0.35	0–2
Influent feed substrate concentration (g l⁻¹)		140	160

and the conditions under which batch, continuous and fed-batch fermentations were carried out are listed in Table 10.1.

10.2.2.2 Measurement of glucose, biomass and product concentration

Off-line glucose was measured using a semi-automatic combined sensor (YSI 27A) consisting of an immobilised glucose oxidase and an oxygen sensor. The on-line glucose analysis was carried out using a fully automated system based on the above combined sensor. Off-line measurements of ethanol concentration were determined using gas chromatography with isopropanol as internal standard while biomass formation was measured turbidimetrically at 620nm as well as dry weight.

10.2.2.3 Measurement of carbon dioxide production rate

The gas output in our fermentation system was found to contain carbon dioxide, oxygen and nitrogen. The concentration of carbon dioxide in the out flow was calculated on the basis of the Gay-lussac stoichiometric equations using a specially designed laboratory micro-flowmeter (Figure 10.3) and the flow rate was found to fall within 1 to 3 $l.h^{-1}$ for batch and continuous processes and between 1 and 32 $l.h^{-1}$ for fed-batch process. The operating principle was to measure the time the gas took to push a given volume of water (9 ml for batch or continuous process and 67.5 ml for fed-batch process) under constant input of air flow containing mainly oxygen and nitrogen.

10.2.2.4 Control and monitoring of fermentation

Monitoring and control of fermentation in our system were achieved through the use of dedicated software and a supervisor.

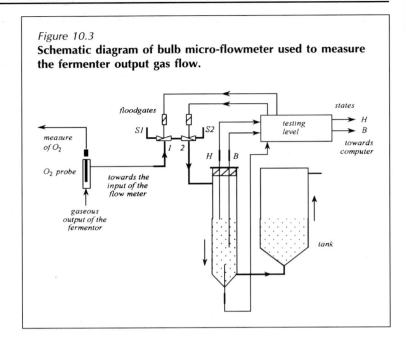

Figure 10.3
Schematic diagram of bulb micro-flowmeter used to measure the fermenter output gas flow.

Functions assumed by the software included: glucose sensor control; dioxide carbon sensor control; data acquisition and storage; graphic display; log book printing; actuator management; numerical application of algorithm, and evaluation of the estimated signal.

10.2.3   *Process modelling*

Process modelling relies very heavily on the kinetics of the reactions involved in the process and as such simulation experiments would have to be carried out in order to develop a model which accurately describes the dynamics of the fermentation process under consideration. A number of models have been developed for the measurement of microbial growth during fermentation, though the model developed by Monod is the most widely used. Microbial growth is usually described by a set of non-linear equations derived from mass-balance and state parameter considerations. The equations described below were derived and formulated on the basis of the 'state space model' which takes into account the liquid and gas phase balances during fermentation. In this mathematical treatment, it was assumed that the fermentation broth is well mixed, i.e. homogeneous. The following set of equations, the observer, describe the fed-batch fermentation process:

Table 10.2 Expression of the different kinetic models for batch, continuous and fed-batch processes

	Specific growth rate $\mu(t)$	Specific substrate consumption rate $v_s(t)$	Specific product consumption rate $v_p(t)$	
Batch process	$\left(\dfrac{\mu_m s(t)}{K_s + s(t)}\left(1 - \dfrac{p(t)}{P_i}\right)\right)$	$\dfrac{\mu(t)}{Y_{X/S}}$	$\dfrac{\mu(t)}{Y_{X/P}}$	
Continuous process	$\mu_m \exp(-k_p P(t))$	$\dfrac{v_p(t)}{Y_{P/S}}$	$\begin{cases} v_m & \text{if } p(t) < 23\ \text{g}_p\text{l}^{-1} \\ \left(\left(v_c + v_m\right)\left(1 - \dfrac{p(t)}{P_i}\right)\right) & \text{otherwise} \end{cases}$	
Fed-batch process	$\left(\dfrac{\mu_m s(t)}{K_s + s(t) + s^2(t)/K_i}\left(1 + \dfrac{p(t)}{P_i}\right)\right)$	$\dfrac{v_p(t)}{Y_{P/S}}$	$\left(\dfrac{v_m s(t)}{K_s' + s(t) + s^2(t)/K_i'}\right)$	

$$\begin{cases} \dfrac{dx(t)}{dt} = \left(\dfrac{\mu_m s(t)}{K_s + s(t)}\left(1 - \dfrac{p(t)}{P_i}\right)\right)x(t) - D(t)x(t) \\[2mm] \dfrac{ds(t)}{dt} = \dfrac{1}{Y_{x/s}}\left(\dfrac{\mu_m s(t)}{K_s + s(t)}\left(1 - \dfrac{p(t)}{P_i}\right)\right)x(t) - D(t)(S_{in}(t) - s(t)) \quad (10.1) \\[2mm] \dfrac{dp(t)}{dt} = \dfrac{1}{Y_{x/p}}\left(\dfrac{\mu_m s(t)}{K_s + s(t)}\left(1 - \dfrac{p(t)}{P_i}\right)\right)x(t) - D(t)p(t) \end{cases}$$

The variables listed above vary from one system of fermentation to another (see Tables 10.2 and 10.3). For example, in the case of a continuous-flow fermentation, the volume $v(t)$ is constant and as such the last equation vanishes and the ratio reduces to the dilution rate. On the other hand, in the case of batch processes, the volume $V(t)$ is constant as there is no feeding; consequently, the last equation vanishes and all the terms representing flow rate become zero. The kinetic models of various modes of operation, the data obtained and the experimental conditions are described in Tables 10.2 and 10.3 respectively.

In a monovariable case under continuous cultivation, the objective of maintaining a constant concentration of substrate within the fermentation vessel is achieved by rigorously controlling the dilution rate $D(t)$. While multivariable, non-linear adaptive control has the main objective of maintaining constant concentrations of both substrate and biomass, the linear adaptive control has the objective of maintaining a constant concentration of substrate as well as a constant rate of carbon dioxide evolution. The latter parameter has proved useful in calculating the rate of

Table 10.3 Data from off-line identified parameters of the reaction models for batch, continuous and fed-batch cultures

	Parameters			
Batch process	$K_s = 5$ g$_s$ l^{-1} $Y_{x/p} = 0.16$ g$_x$ g$_p^{-1}$	$P_1 = 100$ g$_p$ l^{-1}	$\mu_m = 0.45$ h^{-1}	$Y_{x/s} = 0.07$ g$_x$ g$_s^{-1}$
Continuous process	$P_i = 87$ g$_p$ l^{-1} $\nu_m = 0.9$ h^{-1}	$k_p = 0.054$ g$_p^{-1}$ $Y_{p/s} = 0.43$ g$_p$ g$_s^{-1}$	$\mu_m = 0.45$ h^{-1}	$\nu_c = 0.3$ h^{-1}
Fed-batch process	$K_s = 5$ gs l^{-1} $P_i = 100$ g$_p$ l^{-1}	$K'_s = 9$ g$_s$ l^{-1} $\mu_m = 0.54$ h^{-1}	$K_i = 201$ g$_s$ l^{-1} $\nu_m = 2.1$ h^{-1}	$K'_i = 297$ g$_s$ l^{-1} $Y_{p/s} = 0.43$ g$_p$ g$_s^{-1}$

ethanol formation during the course of alcohol fermentation processes. In the multivariable case, the two control variables are the dilution rate $D(t)$ and the substrate concentration in the feed stream $S_{in}(t)$.

10.2.4   Control algorithms

10.2.4.1   Linear monovariable algorithms

The dynamics of alcohol fermentation processes can be appropriately approximated, around their steady state values, by an incremental model representation (Dahhou *et al.*, 1992), with the control objective being achieved though the use of 'partial state model reference control strategy', which can be described mathematically (Saad and Sanchez, 1992) by the following equations:

$$\begin{cases} J = \sum_{j=h}^{h_p}\left\{(e_y(k + j))^2 + \delta(k)(\Delta(q^{-1})e_u(k + j - d))^2\right\} \\ \Delta(q^{-1})e_u(k + j) = 0 \quad \text{for } h_c < j < h_p \end{cases} \tag{10.2}$$

Where the sequences $\{e_u(k)\}$ and $\{e_y(k)\}$ represent the input- and output-tracking errors when the plant model zeros should be preserved in closed-loop, $\delta(k)$ corresponds to the weighting parameter, and h_i, h_c and h_p are the initialization, the control and the prediction horizons respectively. The control law is carried out using the generalized predictive control approach (Clarke *et al.*, 1987). After computing the j-step and the error signal $e_y(k)$, the control law was formulated and expressed as follows:

$$\Delta(q^{-1})e_u(k) = (-1\ 0\ L\ 0)(G^T G + \delta(k)l)^{-1}G^T E_y \tag{10.3}$$

The vector E_y contains the inputs' and outputs' past values, the matrix G is constituted of the impulsional responses of the plant.

10.2.4.2   Linear multivariable algorithms

The linear quadratic control was implemented using the incremental plant parameterization approach (Dahhou *et al.*, 1992). In this system, the regulator is based on the minimization of the following quadratic function:

$$J = \lim_{T \to \infty} E \left\{ \sum_{j=1}^{T} y_c^T(t + j)y_c(t + j) + u_c^T(t + j - 1)\Omega u_c(t + j - 1) \right\}$$

$$(10.4)$$

In this, the cost function Ω is a 2×2 weighting positive matrix. The variables $y_c(t)$ and $u_c(t)$ are defined respectively as $(y(t) - y_r(t))$ and $(u(t) - u(t - 1))$, while $y_r(t)$ is the reference sequence vector generated according to the model previously described (Maher *et al.*, 1995). The control law governing the above control objective is given by:

$$u_c(t) = -K(t)\hat{X}(t) + G(t)(y_c(t) - C_0\hat{X}(t)) \qquad (10.5)$$

The two gain matrices $K(t)$ and $G(t)$ are obtained after the resolution of an algebraic Riccati equation (Maher *et al.*, 1995), while the term $\hat{X}(t)$ represents the estimated states.

10.2.4.3   Non-linear monovariable algorithms

Unlike the algorithms presented in the previous two subsections, the adaptive controllers utilize non-linear equations to describe the non-linear structure of fermentation processes. In this study we have designed an adaptive pole placement program for the control and the regulation of substrate concentration, $s^*(k)$ in the fermentation vessel, by regulating the dilution rate according to the following formula (Dahhou *et al.*, 1993):

$$s(k + 1) = a(k)s(k) + b(k)U(k) \qquad (10.6)$$

We have employed a control mechanism in which the polynomial T_R is an asymptotically stable polynomial whose zeros will be the closed-loop-system poles according to the following formula:

$$\begin{cases} b(k)U(k) = T_R(q^{-1})s^*(k) + (a(k) - T_P(q^{-1}))s(k) \\ T_R(q^{-1}) = 1 + q^{-1}T_P(q^{-1}) \end{cases} \qquad (10.7)$$

On-line estimation of the substrate consumption rate, represented above by the terms $a(k)$ and $b(k)$, was carried out using an adaptive filtering algorithm (Dahhou *et al.*, 1991).

10.2.4.4   Non-linear multivariable algorithms

The control objective here is to minimize the error between the predicted outputs $x^p(k + 2)$, $s^p(k + 2)$ and the reference model outputs $x^{MdR}(k + 2)$, $s^{MdR}(k + 2)$. The two-step ahead predictions of the biomass and substrate concentrations (Queinnec *et al.*, 1991) were derived from the mass-balance equations (10.1) (Euler formula with sampling period T):

$$
\begin{cases}
x^P(k + 2) = x^P(k + 1) + T\mu^P(k + 1)x^P(k + 1) - TD(k + 1)x^P(k + 1) \\
s^P(k + 2) = s^P(k + 1) + \dfrac{T}{Y_{x/s}}\mu^P(k + 1)x^P(k + 1) \\
\qquad\qquad -TD(k + 1)s^P(k + 1) + TD(k + 1)S_{in}(k)
\end{cases}
$$

$$(10.8)$$

The terms $x^p(k + 1)$ and $s^p(k + 1)$, a function of the unmeasured variable $x(t)$ and the unknown specific growth rate $\mu(t)$, were reconstructed with a 'software sensors' in order to minimize the following quadratic cost function with respect to the control variables $D(t)$ and $S_{in}(t)$:

$$
\begin{cases}
\underset{S_{in}(k)}{\text{Min}} \quad J_s = \left\{ s^P(k + 2) - s^{MdR}(k + 2) \right\}^2 \\
S_{in}(k) = S_{in_{min}} + i\dfrac{S_{in_{max}} - S_{in_{min}}}{n_s} \quad i = \text{OL } n_s
\end{cases}
$$

and (10.9)

$$
\begin{cases}
\underset{D(k)}{\text{Min}} \quad J_x = \left\{ x^P(k + 2) - x^{MdR}(k + 2) \right\}^2 \\
D(k) = D_{min} + i\dfrac{D_{min} - D_{max}}{n_x} \quad i = \text{OL } n_x
\end{cases}
$$

where n_x and n_s correspond to the partition number of the admissible control interval [16].

10.2.5 *Results*

10.2.5.1 *Applied control variables*

Real-life experiments using continuous stirred tank bioreactor were conducted to evaluate the performance of the different adaptive controllers.

In the monovariable case the input $D(t)$ was applied directly, while in the multivariable case the inputs $D(t)$ and $S_{in}(t)$ were applied through auxiliary controls (Queinnec *et al.*, 1991). $D(t)$ corresponds to the total of two specific flow rates $D_1(t)$ and $D_2(t)$. $D_1(t)$ represents the specific flow rate of a solution without glucose. $D_2(t)$ corresponds to the specific substrate feeding rate from a solution with a maximal glucose concentration. The following equation describes the relationships among $D(t)$, $S_{in}(t)$, $D_1(t)$ and $D_2(t)$:

$$\begin{cases} D(t) = D_1(t) + D_s(t) \\ S_{in}(t) = \dfrac{D_s(t)}{D_1(t) + D_2(t)} S_{in_{max}}(t) \end{cases} \qquad (10.10)$$

10.2.5.2 Experimental results

Now we will compare the performances of the different control structures presented above. As can be seen, all controllers were first tested systematically and a number of parameters, those which proved satisfactory, were selected. In the linear case, low order models were adopted to approximate the process response, while in the non-linear case, the controller was designed on the bases of the model employed for the process.

The results reported in Figure 10.4 concern the monovariable linear adaptive case. Subplot (*a*) shows the dynamic behaviour of the substrate concentration. After the setpoint value was reached, the steady state was perturbed. As can be seen from Figure 10.4(*b*), the dilution rate, the variable input, responded adequately to the changes that were introduced. The elements of the estimated transfer function shown in subplot (*c*) did not vary when the setpoint was changed from $4g.l^{-1}$ to $2g.l^{-1}$, suggesting that the model is operationally valid within this range.

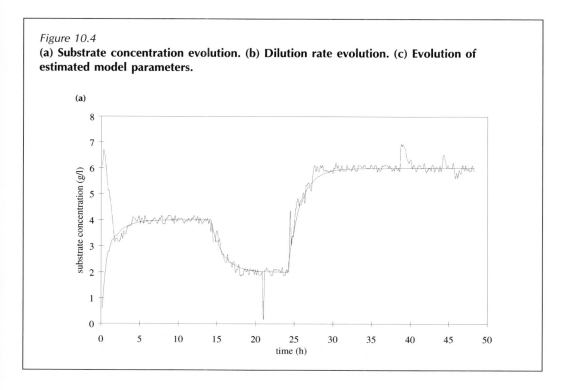

Figure 10.4

(a) Substrate concentration evolution. (b) Dilution rate evolution. (c) Evolution of estimated model parameters.

(a)

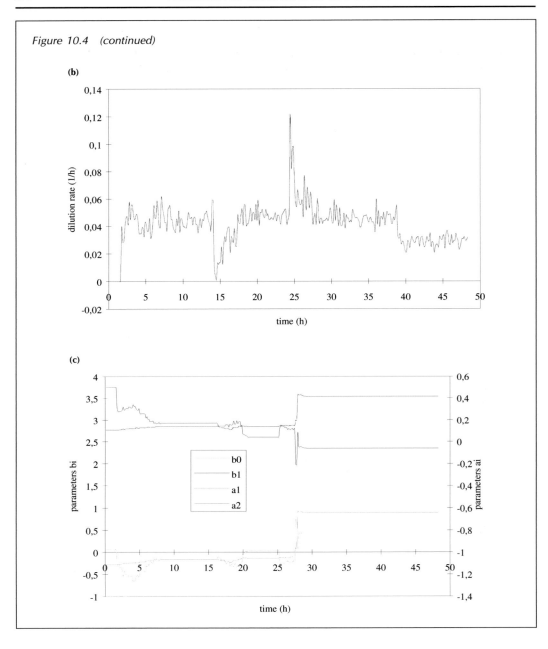

Figure 10.4   (continued)

On the other hand, a change in the estimated model was demonstrated when the setpoint changed from $2g.l^{-1}$ to $7g.l^{-1}$.

The results concerning the multivariable linear adaptive case are shown in Figure 10.5. Here, the controlled variables are the substrate concentration (a) and the carbon dioxide production rate (b). As can be seen from these plots, the controller was able to achieve the two desired setpoints after a short period of time,

Figure 10.5
(a) Substrate concentration evolution. (b) Carbon dioxide production rate evolution.
(c) Evolution of the different dilution rate. (d) Evolution of the different input variables.
(e)–(h) Estimated model parameters.

(a)

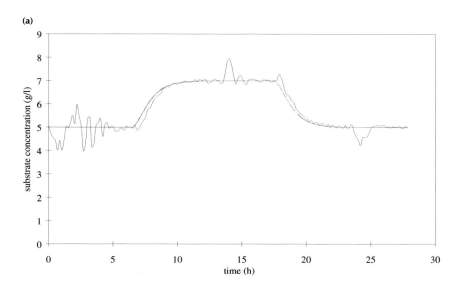

(b)

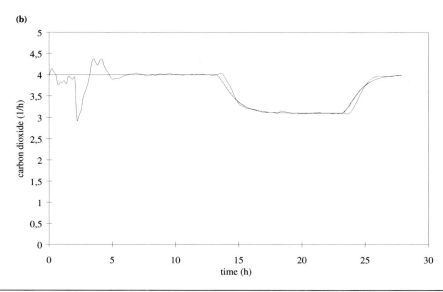

Figure 10.5   (continued)

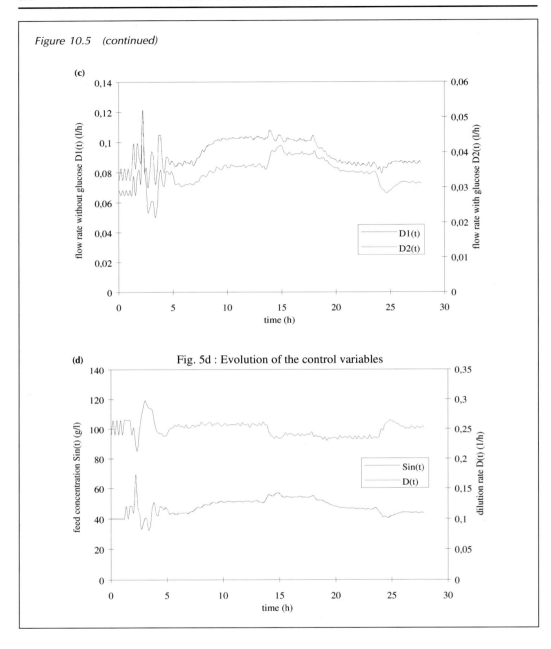

Fig. 5d : Evolution of the control variables

approximately 5 hours. A coupling effect can be seen on the substrate concentration profile at time $t = 13h$ and $t = 23h$ in response to changing the setpoint.

In our experiment, no coupling effect was observed in the carbon dioxide production rate as this parameter is less sensitive than the substrate concentration. Subplots (d) and (c) show the calculated control variables and the applied input variables re-

Figure 10.5   (continued)

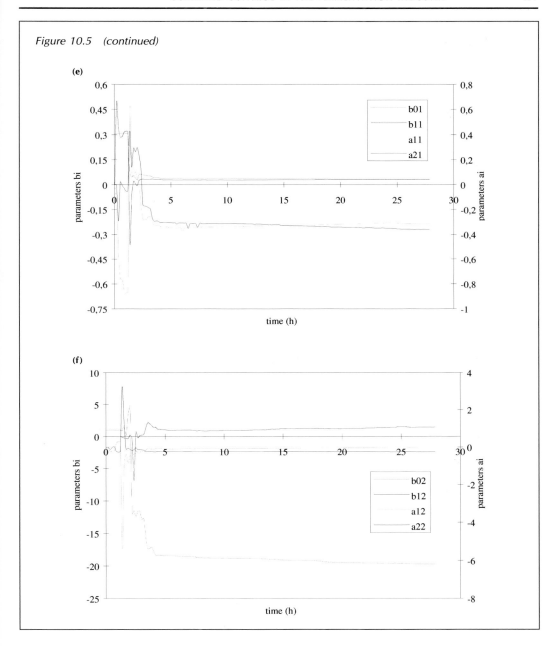

spectively. These variables were very smooth during the course of the experiment, and were never fully saturated. The individual elements of matrix transfer function representing the effect of each input on each output are shown in subplots (*e*), (*f*), (*g*) and (*h*). After four hours, the parameters reached their desired steady state values.

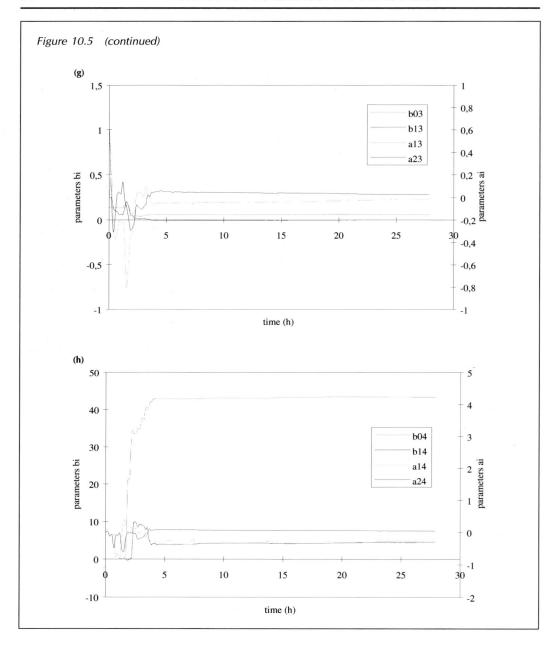

Figure 10.5   (continued)

The data obtained using a monovariable non-linear adaptive control (Figures 10.6a, b and c) appear to be essentially the same as those obtained using the monovariable linear adaptive control. To test the algorithm's capacity to handle perturbation during the course of fermentation, a deliberate cut in the power supply for a few hours was introduced. The data obtained clearly demonstrate that the control algorithm reacted responsively and achieved the steady state set points within a very short space of time (Figure 10.6a). As can seen from subplot (c), the substrate consumption rate was reinitialized at zero following power cuts (at time $t = 14h$ and $t = 22h$) which is contrary to substrate concentration (a) and dilution rate (b) which were reinitialized at the values preceding the power cut.

In another set of experiments, we have adopted a multivariable non-linear adaptive control system in which the regulated variables were the biomass and substrate concentration, shown in subplots (a) and (b) respectively. In this combined approach, all attainable parameters were adjusted so that the algorithm responds efficiently to any changes in the system.

The different control configurations based on a linear or a non-linear adaptive approach gave satisfactory performances for the required control specifications. In the non-linear multivariable case, the performance and decoupling can be improved by the introduction of a penalty on the input and output controls.

In our studies, we have demonstrated the utility of the linear adaptive control (monovariable) and the linear quadratic control (multivariable) during fermentation in a stirred tank bioreactor. The results show that in as far as the input/output behaviour is concerned, the linear adaptive approach is an

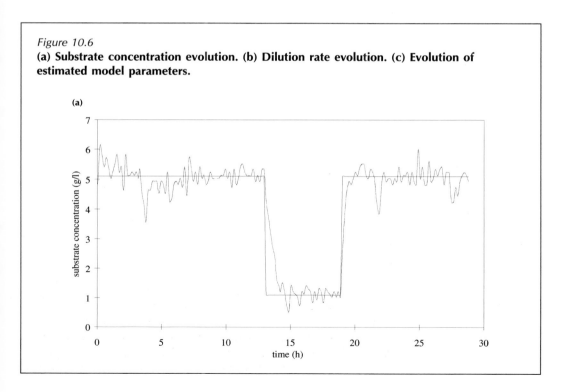

Figure 10.6
(a) Substrate concentration evolution. (b) Dilution rate evolution. (c) Evolution of estimated model parameters.

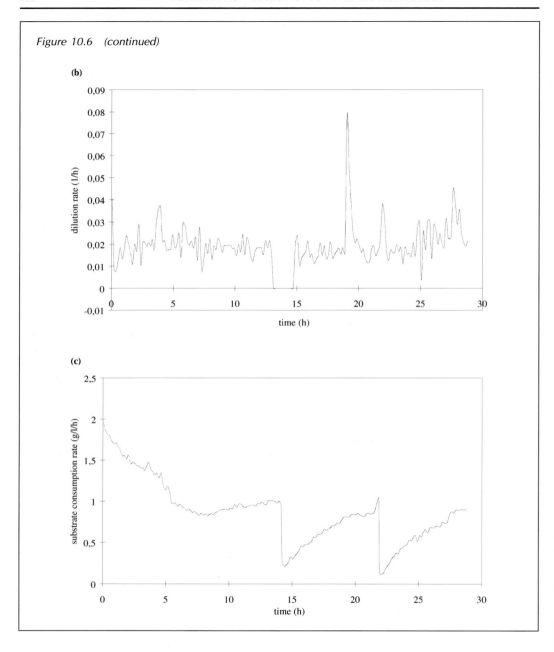

Figure 10.6   (continued)

efficient method. When the algorithms employed for the linear adaptive controls were compared with those formulated and adopted for the non-linear adaptive control used in this study, it was clear that an adaptive control policy based on the non-linear and non-stationary structure of the fermentation process would be more advantageous. Such a strategy provides not only real time information on the culture physiology, but also it reduces the risk of unpleasant on-line errors that often occur when an adaptive linear black box approach is used.

The comprehensive management of a fermentation process requires more than simply an algorithmic approach. 'Expert' and 'Knowledge Based' Systems (KBS), as replacement experts, may not be the complete answer, but the artificial intelligence (AI) techniques recently developed offer a whole host of different strategies which may be adopted in the fermentation industries as well as others.

10.3 Expert control systems and fuzzy logic

10.3.1 Problem statement

The fact that bioprocesses depend on the wide-ranging activities of living organisms makes the task of exercising control and prediction of events during the course of fermentation very difficult indeed. Bioengineers are therefore concerned with devising ways by which microbiologists can control the fermentation process. Moreover, the engineers must also take into account that very little is known about the fine details of the various activities which bring about the formation of the desired end product. In an 'ideal' world, one should be able to have sensors directly inside the cells (see Figure 10.7.a), but unfortunately, at least for the time being, this is not possible. Currently, however, our knowledge of fermentation processes is essentially derived from the wide-ranging variety of data of external parameters, e.g. pO_2, pH, CO_2 etc. Monitoring, control and diagnosis of external parameters are therefore the main tasks which need to be achieved (see Figure 10.7).

Figure 10.7
(a) The 'ideal' solution to the monitoring problem. (b) The current solution to the monitoring problem.

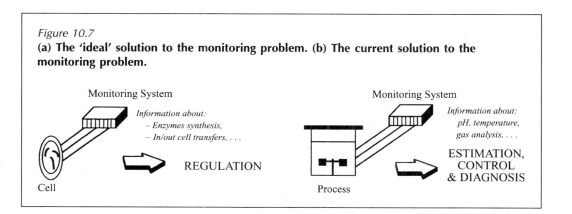

10.3.2   Control of bioprocesses

'Fuzzy logic' is a powerful tool in the control of bioprocesses as it facilitates the incorporation of semi-quantitative and qualitative information into the operating model thus rendering the system simpler and more capable of capturing the human experience. The development of 'fuzzy logic' and its implementation is gathering momentum particularly in the field of wastewater treatment where complex ecosystems are involved. Under similar conditions, a fuzzy controller was capable of providing adequate control over the process with a single conventional strategy. Moreover, intelligent control systems incorporating 'fuzzy logic', expert and hybrid systems which were based on genetic algorithm and neural networks proved successful in the optimization and control of hydroponics growth (Morimoto et al., 1996).

Generally speaking, 'fuzzy logic' as well as all other control systems (quantitative and qualitative) are designed to fulfil three main objectives:

1. to optimize the process (i.e. improve the performances, to keep the process stable despite disturbances);
2. to help the human operator in avoiding drudgeries;
3. to facilitate advanced monitoring, fault detection and, if at all possible, diagnosis.

10.3.3   Fault detection

Fault detection can be achieved using a model-based approach. Research in this field started in the early 1970s with some results on observer-based fault detection in linear systems (Himmelblau, 1978). At the same time, instrument failure detection based on analytical redundancy of multiple observers was demonstrated (Clark, 1978). Although a number of researchers have emphasized the growing interest in using artificial neural networks or fuzzy logic for fault detection purposes, it is our belief that a combined approach (i.e. a combination of more than one approach) could provide better control, as has recently been reported (Steyer et al., 1997) for fault detection in a biological wastewater treatment process (Figure 10.8).

*   Figure 10.8(a) illustrates the system's fault-finding response to deliberate perturbation in the flow rate of feed supply. The objective is to detect these disturbances and respond to them effectively.
*   Figure 10.8(b) shows the output flow rate of gases, the variable parameter to be analysed in order to detect any perturbation in the flow rate of feed supply.

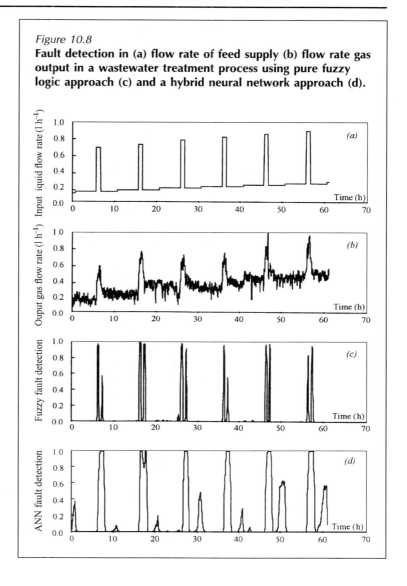

Figure 10.8

Fault detection in (a) flow rate of feed supply (b) flow rate gas output in a wastewater treatment process using pure fuzzy logic approach (c) and a hybrid neural network approach (d).

- Figure 10.8(c) shows the fault detection profile recorded by fuzzy logic based program as it detects changes in the flow rate of the gas output.
- Figure 10.8(d) shows the fault detection profile recorded using a combined approach in which fuzzy qualification of gas output and artificial neural network (ANN) are integrated.

In addition, we would like to point out that variables such as temperature, pH, dissolved oxygen, flow rates – because they are regulated to local setpoints – are not always sufficient to prevent failure. From a fault detection viewpoint, it should be

Figure 10.9
Improvements in increasing the 'relevance' of a measurement by analysing the actuator evolution. (a) Fault detection based on pH values (e.g. threshold comparison). (b) Fault detection based on the frequency of sodium hydroxide addition.

(a)   Time
Fault is detected

(b)   *Time when the pump is turned on for pH regulation*
                                                                  Time
             Difference of time during fault detection

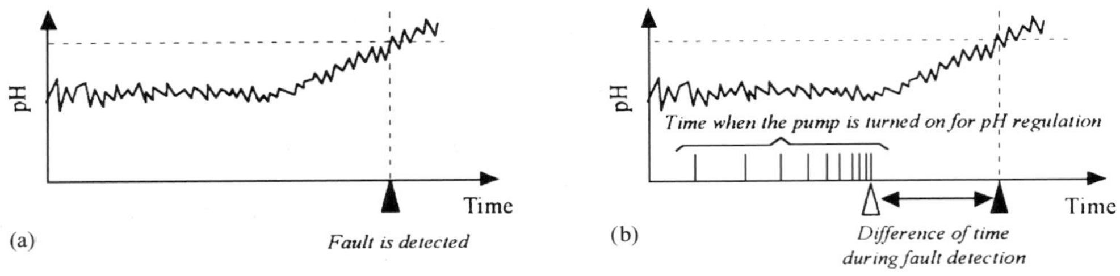

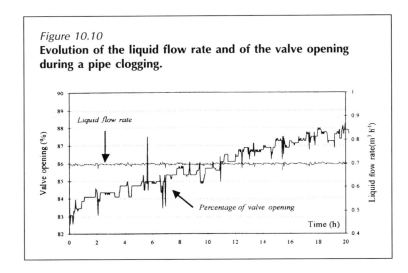

Figure 10.10
Evolution of the liquid flow rate and of the valve opening during a pipe clogging.

remembered that it is very often more rewarding to check the actuator evolution rather than the regulated variable. As an example, it can be very easily demonstrated, by taking into account the frequency of sodium hydroxide addition instead of pH measurements alone, that we can anticipate and minimize malfunctioning in that direction, thus improving the monitoring of the process (see Figure 10.9). In cases where flow rate is regulated through valves, clogging can be avoided more readily by monitoring and analysing valve openings rather than flow rate measurements (see Figure 10.10). In this case, pipe clogging can be detected as soon as the valve opening has increased (in a non-faulty situation, the valve opening is constant) instead of having to wait until a decrease in flow rate is detected (i.e. when the valve opening reaches 100 per cent).

10.3.4 *Fault analysis*

Fault detection is only one of many tasks that need to be considered when developing an advanced monitoring system. Generally speaking, advanced monitoring is concerned with two main aspects: the first is handling of data in real time, while the second is primarily devoted to collecting and displaying data in a meaningful way. To fulfil the above aspects, the following additional tasks have to be carried out:

- to detect, as early as possible, any deviations from normal operating conditions;
- to analyse causes for alarms and to highlight the primary cause without displaying other alarms;
- to indicate recovery steps and the scale of urgency;
- to store normal and abnormal operating conditions that have occurred in the past and to use this process history to guide the operator by giving suggestions on how to improve the current situation.

The introduction of a biological dimension into fuzzy logic and other artificial intelligence systems is steadily increasing and a good number of studies have already been carried out (e.g. Kishimoto *et al.*, 1991, Pokkinen *et al.*, 1992, Kishimoto *et al.*, 1995, Roca *et al.*, 1996), with the view of converting qualitative symbolic expressions into quantitative information so that decisions and actions can be made.

An interesting application of artificial intelligence in bioprocesses is the 'physiological state control' (Konstandinov *et al.*, 1989), which has the main advantage of not requiring conventional mathematical models for the synthesis of the control system. The on-line functions of this system include: (1) the calculation of physiological state variables; (2) the determination of the current physiological situation; and (3) calculation of the control response and actions. The use of 'artificial intelligence' based method, 'fuzzy logic' and 'pattern recognition' theories for the control of phenylalanine production using fed-batch fermentation processes has been reported (Konstandinov *et al.*, 1992). More recently, another study was conducted to handle the biological state of a fermentation process through qualitative rules (Steyer *et al.*, 1993; Steyer *et al.*, 1996). The basic intention was there to build a computer aided-diagnosis rule-based expert system, named BIOTECH, with two main objectives: to explain any event occurring during the fermentation process; and to be able to correct any fluctuation or malfunction in the process.

In order to achieve the above objectives, a general model applicable to many bioprocesses was elaborated with causal graph support as a complementary tool to mathematical models (see

Figure 10.11
Causal graph of a single microorganism – single substrate bioprocess.

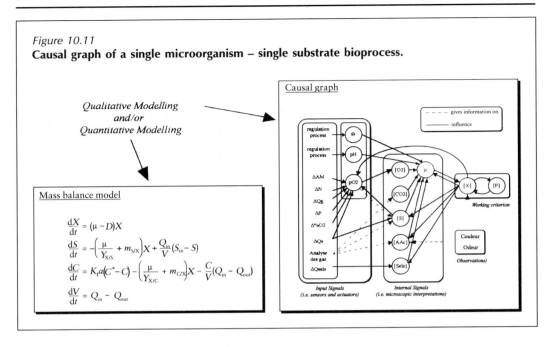

Figure 10.11). This approach has been applied successfully to industrial processes in which bacteria (i.e. *Escherichia coli*) or yeast (i.e. *Saccharomyces cerevisiae*) was used. Under these conditions, more than 35 measurements were made and treated in real-time in order to achieve the following:

- manage sensors faults (meta-rules were introduced to classify the problems according to their consequences)
- choose the best action to be done and, more interestingly, the best control strategy to be used (e.g. which control variable is to be manipulated during the process?)
- predict the end of the process in order to anticipate and better schedule the next operations
- analyse the setting of the regulation processes
- determine the level of contamination and assess the risk involved (temporal reasoning)
- explain the decisions taken.

This approach was shown to be applicable to many bioprocesses (Guerrin *et al.*, 1994), including complex anaerobic digestion processes where more than one single species of microorganism together with different substrates are involved. Such a complex process demands an advanced monitoring system which can be achieved through qualitative modelling and reasoning techniques. The 'causal graph' (Figure 10.12), developed for the control of complex anaerobic digestion processes, is one example.

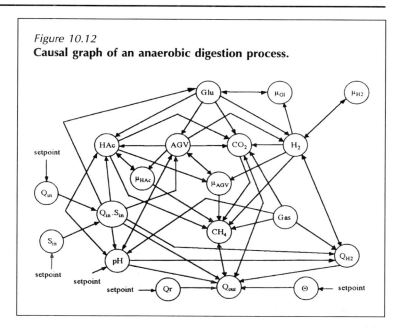

Figure 10.12
Causal graph of an anaerobic digestion process.

These causal graphs can be easily managed with specific qualitative reasoning techniques (Kuipers, 1994). For example, the 'reasoning on the standard state' methodology is a powerful – but simple – approach that is particularly interesting for the on-line diagnosis in bioprocesses. In addition to its simplicity, this approach also affords the following advantages:

- no specific inference engine is necessary and as such it can be implemented by any computer;
- change of the rule base can be made on line;
- evaluation and correspondence between variables are automatically made.

In bioprocesses the measurements of key variables are not always possible and as such the application of conventional monitoring and control theory faces great difficulties. Although the introduction of the so-called 'biological dimension' into control model can lead to a very complex model, recent efforts have been successful in expressing the 'biological dimension' in a qualitative way that is deep enough to describe the microbial activity and simple enough to be used on-line.

10.4   Neural networks

10.4.1   Difficulties in bioprocess control

By and large, the difficulties faced during batch fermentation are due to the lack of adequate on-line measurements, the non-linearity of models and the non-stationary nature of growth dynamics (Royce, 1993). For such processes, the objective is not merely to stabilize the functionality of the system but rather to master the fermentation process from start to finish. Thus, a particular effort has to be undertaken in the modelization of transitory states, a goal which has been achieved through the use of a neural network model system. Models of static and dynamic neural networks are used to estimate (at the considered time), or to predict (future values) values of key variables which are not directly measurable due to the lack of sensors. These variables are generally related to biomass formation (Tiessier *et al.*, 1996 and 1997), fermentation kinetics (Cleeran *et al.*, 1991), concentration of metabolites (Linko *et al.*, 1995) and, to a lesser extent, enzymatic activities (Linko *et al.*, 1997).

The utilization of dynamic neural network models, which take into consideration the discontinuous nature of fermentation processes, allowed accurate prediction of the endpoint of alcohol fermentation in the wine industry (Bochereau *et al.*, 1991), with an acceptable margin of error, less than 13 per cent. Similarly, Latrille *et al.* (1994) have used recurrent neural networks for the prediction of the end point of lactic acid fermentation in yoghurt production. In this case, relevant on-line measurements are fed into the neural models where such information is tested against a set of specific and desirable criteria for the final product. This approach is of great interest to the fermentation industry as well as many others, particularly those where no sensors are available for the measurement of key variables.

10.4.2 *Implementation of neural net and identification of parameters*

Dynamic neural systems which usually consist of two layers of perceptrons, the hidden layer and the output layer, approximate non-linear and dynamic functions with consummate ease. The hidden layer comprises neurons with sigmoidal activation function $f(x) = 1/(1 + \exp(-x))$. In this case, values of output variables are between 0 and 1. A standardization of data is realized by assigning input and output parameters (key variables) values between 0 to 1.0 (minimum and maximum). Such a standardization has the added advantage of achieving uniformity among variables irrespective of their unit of measure or variance. The operations undertaken on each layer can be described by the following equations:

Hidden layer: $1 \leq j \leq J$

$$O_j = f\left[\sum_{i=1}^{I} W_{ij} \cdot X_i + \beta_j \right] \tag{10.11}$$

> *Figure 10.13*
> **The interrelationship between input and output in static (a)
> and dynamic (b) models. X1(*t*) and X2(*t*) are state variables at
> time *t*. U(*t*) is an external or variable. Y1(*t*) and Y2(*t*) are
> measurement variables at time *t*.**
>
>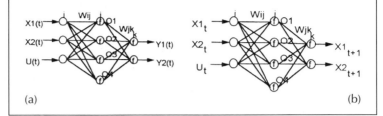
>
> (a)   (b)

Output layer: $1 \leq k \leq K$

$$Y_k = f\left[\sum_{j=1}^{J} W_{jk} \cdot O_j + \beta_{J+k}\right] \tag{10.12}$$

Where $X(X_i)$ is the normalized input, $O(O_j)$ is the output vector
of neurons of the hidden layer and $Y(Y_k)$ is the normalized output vector of the network. While W_{ij} & W_{jk} are weight matrices
connecting adjacent neurons layers, β_i is the bias vector of
neurons. The interrelationship between input and output in static
and dynamic models is highlighted in Figure 10.13(a) and (b)
respectively. In both cases, command variables were introduced
into the input domain of the network.

The implementation of a system of neurons requires clarifica-
tion and full knowledge of the following points.

10.4.2.1   *Choice of the variables*

A good knowledge of the process allows not only the identification of relevant variables but also the definition of the role of
each variable (state, measurement or command) in the model
system. For a dynamic system, it is necessary also to define the
order of each state variable. A variable of order *i* is a variable
whose effect on the process, at instant *t*, depends on the values
recorded for this variable in all preceding instants between $t-1$
and $t-i$.

10.4.2.2   *Selection and distribution of data*

Effective use of non-linear control models requires a comprehensive range of information in order to assess the influence of
each variable on the process. The distribution of data under
these circumstances is generally distributed in two sets:

- a learning set, comprising approximately 70% of the data;
- a test set containing the remaining data.

The distribution of the data in the above sets is carried out either randomly (random drawing) or statistically, dependent on the design employed. A good distribution of data is central to obtaining an effective neural model. A third set, the validation set, may also be constituted to validate the performance of the model without bias.

10.4.2.3   Choice of an algorithm for minimization

The most effective algorithms, for multi-layers perceptrons, are those classically used for the minimization of a quadratic error, e.g. the conjugated gradient and the algorithm of quasi-Newton (Bishop, 1994). The utilization of a method of quasi-Newton necessitates the calculation of the gradient of the error related to the parameters of the model. Multi-layer perceptrons allow an easy calculation of this gradient by using the technique of back propagation of the error (Rumelhart *et al.*, 1986). According to the type of neural network, static or recurrent, the determination of this gradient will be more or less complex. This supposes the choice of a well-adapted method of learning set.

10.4.2.4   Choice of a method of learning

In the case of a state model where state variables are measured, the dynamic equation describing the system can be written as follows:

$$X_{t+1} = f_{NN}(X_t, U_t, p) \tag{10.13}$$

where X represents the state vector, U represents the command vector, f_{NN}, represents a neural model and p represents the model's parameters. For the identification of the function f_{NN}, directed or semi-directed learning methods can be applied (Nerrand *et al.*, 1993).

Learning phase in non-recurrent or directed mode
In the case of non-recurrent or directed mode, the equation (10.13) becomes:

$$\hat{X}_{t+1} = f_{NN}(X_t, U_t, p) \tag{10.14}$$

where the sign $\wedge$ indicates an estimated variable. The absence of this symbol indicates a measured variable. One presents the system (in all instants), with input variables measured at a particular time t and, output variables measured at another point, $t + 1$. The error E, defined by the following relationship (10.15), has to be minimized:

$$E = \sqrt{\frac{\sum_{j=1}^{M}\sum_{i=k}^{n_j-1}(X_{i+1} - f_{NN}(X_i, X_{i-1},\ldots, X_{i-k}, U_i, p))^2}{\sum_{j=1}^{M}(n_j - k)}} \qquad (10.15)$$

In the above equation, M is the number of experiments, n_j is the number of data of each experiment, k is the order of the system and X is the normalized vector of output variables. This method applies to cases where reaction kinetics are not reproducible.

Learning phase in recurrent or semi-directed mode
The recurrent mode corresponds to a semi-directed or parallel learning process in which the model operates only for a certain period of time which corresponds to the chosen horizon of prediction. Under these circumstances, equation (10.13) becomes:

$$\hat{X}_{t+1} = f_{NN}(\hat{X}_t, U_t, p) \qquad (10.16)$$

With a horizon of prediction H and a sliding window width G, such that $G \leq H$, the error E to be minimized is defined by the relationship (10.17):

$$E = \sqrt{\frac{\sum_{j=1}^{M}\sum_{g=1}^{G}\sum_{i=k}^{n_j-H}(X_{i+H-g+1} - f_{NN} \circ f_{NN} \text{ OK } \circ f_{NN}(X_i, X_{i-1},\ldots, X_{i-k}, U_i, p))^2}{\sum_{j=1}^{M}(n_j - k - H + 1)\cdot G}}$$

$$(10.17)$$

Both the horizon of prediction H and the width of the variable window Gm are determined by trials and errors and guided by the dynamics of the system to be modelled.

In practice, once a first directed learning process is realized, the weight values obtained allow the initialization of a semi-directed learning process. It is desirable to initialize the minimization using values close to the minimum so as to avoid the need for further reiteration of numbers.

10.4.2.5   *Practice of the learning process*

During the learning phase, the experienced operator seeks to avoid the following difficulties.

Local minimum
Identification of system parameters for neurons based system is an iterative procedure, a non-linear regression, and as such the parameters values must be close to experimentally determined

values, e.g. using the results obtained from directed learning. Alternatively, each of these parameters (weight of the neural network) can be initialized a random value between −1 and 1.

Over-parametrization of the neural network

Over-parameterization occurs when the number of parameters in a given neural model exceeds available data. Generally speaking, the number of data has to be 5 to 10 times higher than the number of parameters. However, this gives no indication as to what is the minimal number of parameters that have to be used without compromising the 'optimal' structure of the model.

An 'optimal' architecture selection of neural networks structure can, however, be achieved by suppressing certain neurons in the hidden layer with the aid of statistical analysis techniques such as Chi2 (Urbani *et al.*, 1994) or other directly related criteria (MacKay, 1995).

The over-learning of the neural network: utilization of a cross-validation method

Despite the selection of a neural system of reduced size (limited number of neurons in the hidden layer), an over-learning of the model may be observed. It is characterized by a very good approximation of the data in the learning set and a very poor approximation of the data in the test and validation sets. The recording of errors as a function of iteration is essential as it yields values of weights that are directly related to the functionality of the test. This procedure, illustrated in Figure 10.14, minimizes the risk of over-learning of the neural network.

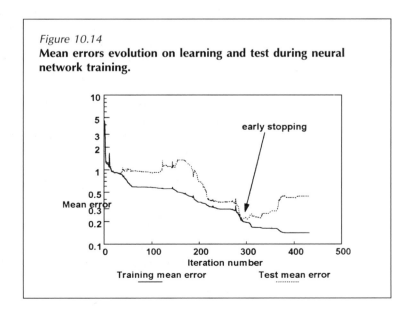

Figure 10.14

Mean errors evolution on learning and test during neural network training.

10.4.2.6 *Validation of the model by simulation*

The validation and, in turn, the final acceptance of a neural model depends on its ability to respond and control biological processes. To proceed, the model will be subjected to a new set of data and the simulation results will give a clear indication of the validity and suitability of the model for the control of a given process.

10.4.3 *Examples of application to lactic fermentations*

Significant progress has been made over the last few years in developing methods for the indirect measurement of biological parameters during the course of lactic acid fermentations in general and the production of starter cultures, fermented milk and fresh cheese in particular. In all lactic acid fermentation processes, the electrical conductivity of the culture medium appears to change as a result of increasing acid production. It must be remembered, however, that electrical conductivity varies with temperature and medium composition. Recently, mathematical models of lactic acid fermentation were developed and integrated into computer programs, taking into consideration the stoichiometry of various reactions in the pathway as well as the influence of media composition and temperature on electrical conductivity (Latrille *et al.*, 1996).

10.4.3.1 *Bacterial biomass*

The production of lactic acid during the course of fermentation may or may not, depending on the organism under investigation, affect growth. The inhibitory effect of lactic acid on growth, however, can be easily assessed through the use of non-linear regression models as has been demonstrated in the case of thermophilic lactofermentors (Acuna *et al.*, 1994). Both dynamic or static models (Figure 10.15) of neural networks are suitable for this kind of work.

From the temperature (T), the pH, the concentration of lactic acid (P) and the initial concentration of bacteria (Xo) they were able to describe the growth pattern of *Streptococcus thermophilus* and *Lactobacillus bulgaricus* during the course of fermentation with acceptable average of experimental errors (between 5 and 15 per cent). Several examples are presented in Figure 10.16.

10.4.3.2 *Indirect measurements and prediction of the pH*

The measurement of the pH is particularly important in the dairy industry as most processes are influenced directly or indirectly by the formation of lactic acid, e.g. the manufacture of yoghurt and fresh cheese. The endpoint of lactic acid fermentation can,

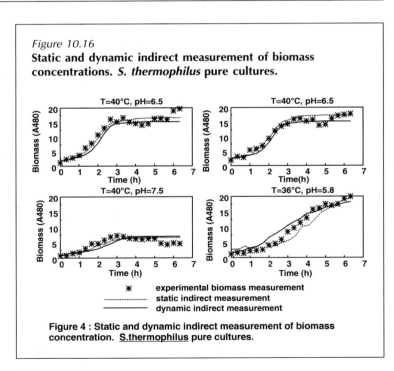

Figure 10.15
Schematic representation of static and neural network models established to determine lactic acid bacteria concentration.

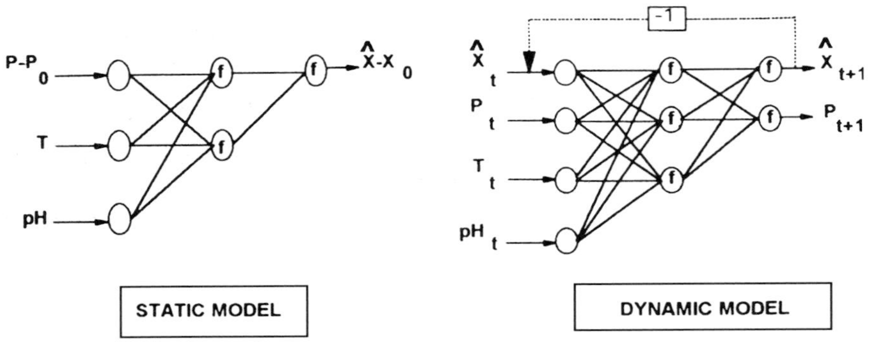

STATIC MODEL

DYNAMIC MODEL

Figure 10.16
Static and dynamic indirect measurement of biomass concentrations. *S. thermophilus* pure cultures.

* experimental biomass measurement
.... static indirect measurement
——— dynamic indirect measurement

Figure 4 : Static and dynamic indirect measurement of biomass concentration. S.thermophilus pure cultures.

therefore, be identified as the pH value at which both high yield and desirable organoleptic qualities are obtained. At the industrial level, however, the measurement of the pH is rarely realized on-line, even if probes are available, due to fouling and failures in calibration. Indirect measurements of pH by following the changes in electrical conductivity is an attractive alternative. Unlike pH probes, electrical conductivity probes are not sensitive to fouling and do not necessitate systematic calibration. A

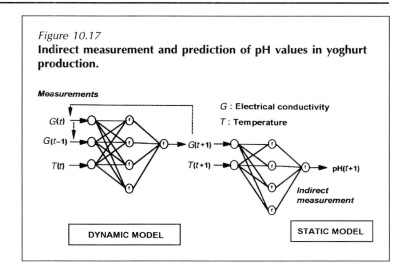

Figure 10.17
Indirect measurement and prediction of pH values in yoghurt production.

Measurements

G(t)
G(t-1)
T(t)

G : Electrical conductivity
T : Temperature

G(t+1)
T(t+1)

pH(t+1)

Indirect measurement

DYNAMIC MODEL

STATIC MODEL

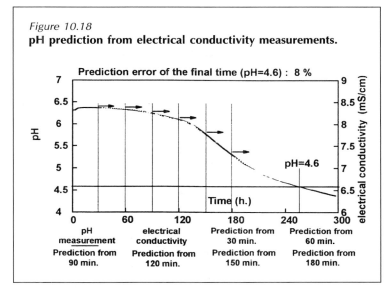

Figure 10.18
pH prediction from electrical conductivity measurements.

Prediction error of the final time (pH=4.6) : 8 %

pH=4.6

Time (h.)

pH
electrical conductivity (mS/cm)

pH
measurement

electrical
conductivity

Prediction from
30 min.

Prediction from
60 min.

Prediction from
90 min.

Prediction from
120 min.

Prediction from
150 min.

Prediction from
180 min.

static model based on a neural network was, therefore, established to determine the pH from electrical conductivity and temperature measurements (Latrille *et al.*, 1996). Similarly, the introduction of a neural dynamic model in association with the static model gives the system an added power of prediction.

Figure 10.17 highlights the architecture of the two neural networks used for the calculation of pH and the determination of the point at which fermentation should be terminated. In this case, the end of the fermentation was identified when the pH has reached a certain value.

Figure 10.18 shows an example that allows the evaluation of indirect measurements of pH as a suitable alternative for monitoring the drop in pH during the course of lactic acid fermentation and for prediction of pH profile as well as the calculation of the point at which fermentation should be terminated. This approach proved successful in the manufacture of fermented milk products and can be equally successful in other products.

Summary

During the course of fermentation processes, the determination of biomass, substrates and metabolites concentrations is central to its success. Generally, these variables are not accessible in real time with the help of sensors. The utilization of indirect measurements and estimation models proved promising in some cases and invaluable in others.

Static models are relatively simple but rather sensitive to fluctuations or perturbations. Dynamic models on the other hand are less sensitive to perturbations and can be incorporated into predictive and diagnostic strategies or advanced control programs.

The use of effective algorithms and powerful computers reduced the time required for the calculation and identification of parameters of a given model. Validation of different control programs, however, requires the availability of a sufficiently rich database of all fermentation parameters and variables.

A good deal of research is currently devoted towards innovation in the following areas:

- mixed neural networks design associating knowledge and neural models of 'black box' type;

- utilization of numerical estimators with sliding horizon;

- the calculation of various parameters, in real time, using an algorithm incorporating the biological dimension.

These technologies should offer, in a reasonable period of time, very efficient methods but their applicability to different fermentation processes would have to be evaluated.

References

Acuña, G., Latille, E., Beal, C., Corrieu, G. and Cheruy, A. (1994) On-line estimation of biological variables during pH controlled lactic acid fermentations, *Biotechnol. Bioeng.*, **44**, 1168–1176.

Bastin, G. and Dochain, D. (1990) *On-line Estimation and Adaptive Control of Bioreactors*, Amsterdam: Elsevier Science Publisher BV.

Bishop, C.M. (1994) Neural networks and their applications, *Rev. Sci. Instrum.*, **65**, 1803–1832.

Bochereau, L., Bourgine, P., Bouyer, F. and Muratet, G. (1991) Modélisation de réacteurs discontinus à l'aide de réseaux connexionnistes, in Antonini, G. and Ben Aim, R. (eds) *Récents progrès en génie des procédés*, Vol. 5, No. 13, Paris: Lavoisier-Technique et Documentation, pp. 385–390.

Clarke, D.W., Mohtadi, C. and Tuffs, P.S. (1987) Generalized predictive control, Parts I and II, *Automatica*, **23**, 137–160.

Cleran, Y., Thibault, J., Cheruy, A. and Corrieu, G. (1991) Comparison of prediction performances between models obtained by the group method of data handling and neural networks for the alcoholic fermentation rate in enology. *J. Ferm. Biotechnol.*, **71**, 356–362.

Dahhou, B., Roux, G. and Cheruy, A. (1993) Linear and nonlinear adaptive control of alcoholic fermentation process: experimental results, *Int. J. of Adaptive Control and Signal Processing*, **7**, 213–233.

Dahhou, B., Roux, G. and Queinnec, I. (1992) Robust adaptive predictive control of biotechnological process: experimental results, Preprints in IFAC Int. Symposium, ACASP '92, pp. 549–554.

Dahhou, B., Roux, G., Queinnec, I. and Pourciel, J.B. (1991) Adaptive pole placement of a continuous fermentation process, *Int. J. System Sci.*, **22**, 2625–2638.

Guerrin, F., Bousson, K., Steyer, J.-Ph. and Travé-Massuyès, L. (1994) Qualitative reasoning methods for CELSS modeling, *Adv. Space Res.*, **14**(11), 307–312.

Himmelblau, D.M. (1978) *Fault Detection and Diagnosis in Chemical and Petrochemical Processes*, Amsterdam: Elsevier.

Kishimoto, M. and Suzuki, H. (1995) Application of an expert system to high cell density cultivation of *Escherichia coli*, *J. Ferment. Bioeng.*, **80**, 58–62.

Kishimoto, M., Moo-Young, M. and Allsop, P. (1991) A fuzzy expert system for the optimization of glutamic acid production, *Bioproc. Eng.*, **6**, 163–172.

Konstandinov, K.B. and Yoshida, T. (1989) Physiological state control of fermentation processes, *Biotechnol. Bioeng.*, **33**, 1145–1156.

Konstandinov, K.B., Matanguihan, R.M. and Yoshida, T. (1992) Physiological state control of recombinant amino acid production using a micro expert system with modular, embedded architecture, in Proceedings of IFAC Modeling and Control of Biotechnical Processes, Colorado, USA, pp. 411–414.

Kuipers, B. (1994) *Qualitative Reasoning: Modeling and Simulation with Incomplete Knowledge*, Artificial Intelligence Series, Cambridge, MA: The MIT Press.

Latrille, E., Acuña, G. and Corrieu, G. (1996) Application des réseaux de neurones pour la modélisation de bioprocédés discontinu. *J. Eur. Des Systèmes Automatisés: Intelligence Artificielle et Automatique*, **30**, 357–379.

Latrille, E., Corrieu, G. and Thibauilt, J. (1994) Neural network models for final process time determination in fermented milk production. *Computers Chem. Eng.*, **18**, 1171–1181.

Linko, S., Loupa, J. and Zhu, Y.H. (1997) Neural network as 'software sensors' in enzyme production. *J. Biotechnol.*, **52**, 257–266.

Linko, S., Rajaalahti, T. and Zhu, Y.H. (1995) Neural state estimation and prediction in amino acid fermentation. *Biotechnol. Tech.*, **9**, 607–612.

MacKay, D.J.C. (1995) Probabilistic networks: new models and new methods, in Fogelman-Soulié, F. and Gallinari, P. (eds) Proceedings of International Conference on Artificial Neural Networks (ICANN '95), Vol. 1, Paris: EC2 & Cie, pp. 331–337.

Maher, M., Dahhou, B. and Roux, G. (1995) Multivariable adaptive control of bioprocess, in Proceedings of 3rd IEEE Mediterranean Intl Symposium, Vol. 1, pp. 238–242.

Pokkinen, M., Flores Bustamante, Z.R., Asama, H., Endo, I., Aarts, R. and Linko, P. (1992) A knowledge based system for diagnosing microbial activities during a fermentation process, *Bioprocess Eng.*, **7**, 331–334.

Queinnec, I. and Dahhou, B. (1994) Optimization and control of a fedbatch fermentation process', *J. Optimal Control Applications and Methods*, **15**, 175–191.

Queinnec, I., Dahhou, B., Roux, G., Goma, G. and Pourciel, J.B. (1991) Estimation and control of a continuous alcoholic fermentation process, *J. Ferm. Bioeng.*, **72/4**, 285–290.

Queinnec, I., Roux, G. and Dahhou, B. (1993) Multivariable adaptive predictive control of an alcoholic fermentation process, in Proceedings of 2nd European Control Conference, ECC '93, Vol. 3, 1736–1740.

Roca, E., Flores, J., Rodriguez, I., Cameselle, C., Nunez, M.J. and Lema, J.M. (1996) Knowledge-based control applied to fixed bed pulsed bioreactor, *Bioproc. Eng.*, **14**, 113–118.

Royce, P. (1993) A discussion of recent developments in fermentation monitoring and control from a practical perspective, *Crit. Rev. Biotechnol.*, **13**, 117–149.

Rumelhart, D.E., Hinton, E. and Williams, R.J. (1986) Learning representations by back-propagating errors. *Nature*, **323**, 533–536.

Saad, M.M. and Sanchez, G. (1992) Partial state reference model adaptive control of multivariable systems, *Automatica*, **28**(6), 1189–1197.

Steyer, J.-Ph., Queinnec, I. and Simoes, D. (1993) Biotech: a real-time application of artificial intelligence for fermentation processes, *Control Engineering Practice*, **1**(2), 315–321.

Steyer, J.-Ph., Queinnec, I., Capit, F. and Pourciel, J.-B. (1996) Qualitative rules as a way to handle the biological state of a fermentation process: an industrial application, *J. Européen des Systèmes Automatisés RAIRO-APII*, **30**(2/3), 381–398.

Steyer, J.-Ph., Rolland, D., Bouvier, J.C. and Moletta, R. (1997) Hybrid fuzzy neural network for diagnosis – application to the anaerobic treatment of wine distillery wastewater in a fluidized bed reactor, *Water Sci. Technol.*, **36**, 209–217.

Teissier, P., Perret, B., Latrille, E., Barillere, J.M. and Corrieu, G. (1996) Yeast concentration estimation and prediction with static and dynamic neural network models in batch cultures, *Bioproc. Eng.*, **14**, 231–235.

Teissier P., Perret, B., Latrille, E., Barillere, J.M. and Corrieu, G. (1997) A hybrid recurrent neural network model for yeast production monitoring and control in a wine base medium. *J. Biotechnol.*, **55**, 157–169.

Urbani, D., Roussel-Ragot, P., Personnaz, L. and Dreyfus, G. (1994) The selection of neural models of non-linear dynamical systems by statistical tests, in Vlontzos, J., Hwang, J. and Wilson, E. (eds) Proceedings of Neural Networks for Signal Processing IV, New York, IEEE Press.

Vigié, P. (1990) Contribution à l'optimisation de la fermentation alcoolique par cultures continues en réacteur cascade, Doctorate thesis, INSA, Toulouse, France.

Index

Note: *italicised* numbers indicate figures, tables or boxes **emboldened** numbers indicate definitions